A Letter About Exposing
a Science Cover-Up

Written by Peet (P.S.J.) Schutte

© KOSMOLOGIESE EN ASTRONOMIESE TEGNIKA

All rights are reserved.
No part, parts or the entirety of this book may be reproduced by publishing, electronically copied, duplicated by whatever means that form reproduction or duplication, without the prior written consent of the copy rite owner.

ISBN-13: 978-1499755787

ISBN-10: 1499755783

My personal advice to any potential reader of my writing: My style needs very careful reading: Do not try to speed read or do glance reading but read every word I use carefully and you will do well in understanding because with my writing style it requires much concentration. I don't waste words and I compress factual information.

The Science Whistleblower

Which is a better informing version of
The Science Informer

This letter has other versions as well that varies in palatability.

Letter choice The Exposing Letter ISBN-13: 978-1499755787 ISBN-10: 1499755783
You are reading The Revealing on a Science Conspiracy

Letter choice The Informing Letter ISBN-13: 978-1500326678   ISBN-10: 1500326674
There is also The Informer of a Science Conspiracy

Letter choice The Revealing Letter ISBN-13: 978-1500326708   ISBN-10: 1500326704
There is also The Exposing of a Science Conspiracy.

I have three versions of precisely the same letter. It deals with science and science has a wide-ranging field of interest. Some people including my wife and my daughter in law are easily convinced and need little information to reach a conclusion whichever way. More information would not change their opinion either way.
I did this categorization with those close to me in mind because we as a family are as ordinary as everybody in the world. Then my sons and my brothers require more information to be informed on any subject as to be convinced but don't like to be overburdened by much explaining and in depth study results.

WHOM IT MAY CONCERN,

I am P.S.J Schutte, nicknamed Peet. Being a white South African my mother tongue is Afrikaans and my second language is English. I have per suiting a new cosmic theory that I partly present in a six part theses, of which the investigating research began in 1977. First I located what was wrong in physics. I compiled my presentation of The theses called The Absolute Relevancy of Singularity and then six separate thesis parts forming the theses published through LULU.com which I saw as way the only manner whereby I could generate funding by which I would be able to have the thirty seven or so books I already wrote linguistically edited and then to have the books published on a Print-On-Demand basis. I compiled a new cosmic theory by which I eliminated all the incorrectness that Newton has burdened science with but with this being my opinion I did not find a garage full of academics supporters waiting to applaud me and to uphold my views on the matter. Yet still I was not going to be ambushed by their relentless stonewalling my efforts and blocking my efforts in introducing both the incorrectness and the new cosmic theorem I concluded. Their blocking convinced me about a Conspiracy in Science in Progress and this spurred me on to tell the entire world about their brainwashing of the minds of students.

This kept me busy for the past going on to twelve years on full time basis whereby I was trying to introduce my findings to many academics without finding much joy from my efforts. This past almost thirteen years plus saw me go without any income as I tried to get my theorem recognised as well as get my warning noted. Going without a steady income left me almost destitute and in order to find a manner to get my theory across to the attention of influential readers, I decided to publish a theses of six books electronically as to try and get around the stranglehold of Newtonian bias controlling science at present worldwide. I decided to publish electronically which those in power do not control.

It is said that when any person is capable of understanding 500 words in any language such a person is able to converse in that language. Well I have not counted the exact number of words that I can use to express myself in English but my guess is that it should be close to 495 words that I have mastered in English and so by reading this book you then would be a witness to this. With my first language not English and the books not linguistically checked by an expert there are bound to be language errors that readers will notice. In the past I tried to check my work myself but after checking say one hundred and fifty pages for language corrections, then after days of toiling instead of having corrected work I ended having four hundred pages of newly written information which is still not linguistically corrected but holds a lot more information. The language and spelling errors compiled instead of reduced. This is because my priorities lie elsewhere. I aim to spend money on correcting the work as far as language goes, as I receive money in the selling of my theses and in the hope that I will receive money. I will have all my work including the one you are reading edited professionally and corrected as I find money to do so...But first I have to get the public aware of the problem to get the academics to appreciate the problem.

This is my introduction and this is my prologue: But before I can commence with that task I have another duty to administer: I am about to warn every person in sight of my work about my preserved slender abilities...pointing out weaknesses seen by experts and it is in response to these acquisitions about my approach not matching the normal norm and by me not conforming to the current accepted norms I am expected to confirm their believes never taken with equanimity. The current norms are correct. Any deviation from such thoughts is unacceptable.

Therefore in the light of what the most respected academic group on Earth accuses me of, I therefore have to issue a most serious warning to any person with the intention of making some kind of inquiry to the content this book holds, then the most concerning matter involving any content within the pages of this book you inquire to acquire that you must please seriously consider that where the stating declares the possibility that the content this book has been (written by...) then don't take the announcing Written By Peet Schutte (Petrus S. J. Schutte very seriously for there are grievous doubts leaving considerable dispute about the possibility, which underwrites the authenticity of Peet Schutte achieving the (written by...Peet Schutte status. Please take note of the following dehortation In the light of the reference to me serving in the capacity as being responsible for authoring, (written by   in line of keeping fairness and justice to members of society, where all civil beings should carry reputed honesty, then: Please be warned before any reader starts reading about the following extremely serious admonition: I am bound by my conscience to warn all intended readers that I am placed under caution by the Academics in Physics. Those most esteemed members responsible for the guardianship and maintaining the ethos in physics are of the opinion that I, Peet Schutte, am unable to write any book on the science of Physics as well as Astrophysics. Therefore, I, Peet Schutte, must declare that I should be considered as not very able to write anything, because I am incapable thereof. I suppose, I merely generate new information, which I establish as thoughts and then gather as concepts. I further collect the result as words, which I put on paper using alphabetic symbols. I then compile that in a format that others may confuse with a book, but a book it cannot be, since the Masters in science found me unable to write a book. However, as you are about to see in due course that I am warned by the most esteemed academics in physics that I cannot use words to describe physics. But before you go further and follow my arguments, I first have to level with you about how academics view me in the position I hold. Please do not allow me to fool you, for this then cannot be, or represent a book because I use words and words are what Astrophysicists don't use because then they might detect their stupidity.

I didn't write any books since according to those wise enough to judge me without ever reading my work and form an opinion without being informed about my work's content that I am not schooled to do so. It is my guess that I merely generated uninformed thoughts, which I collected as alphabetic symbols and plotted that in ink on paper. This effort I achieved from harbouring my delusional ideas spawned by a dehumanised brain. It only proves my weak and under developed mentality, due to my lack of an informed insight that is a typical symptom that all those have that is suffering from a disadvantaged past that one can only have when the person obviously lacks a Newtonian opinion. While you are reading the letter deciding to regard or dismiss my work, then also please keep in mind when reading my language used and also please give credit where it belongs…if you do find linguistically improper use of words or misspelling, then remember that I am a feeble minded because I disagree with Newton and not a literal giant. Now I have done my duty in warning everyone and in that, I denounce further participating with any purposive intention to wilfully bring down the crux of civilization by acting unacceptable and irresponsible.

I do find much pride in my status as being Afrikaner and would like to have my names used by pronouncing it in the manner Afrikaans dictates…therefore I would sincerely appreciate the courtesy when readers will take note that my name and last name are pronounced in Afrikaans, which is originally from Dutch and must be pronounced that way. Peet one would pronounce "here" which is the closest English to the pronouncing of the "ee". The "Sch" in Schutte is pronounced exactly as school is where both actually are pronounced Skutte or "skool". By pronouncing my name in Afrikaans you do me the utmost courtesy any one can. Being an Afrikaner is what I am most proud of. I submit article to well known physics magazines but my articles are rejected on the most unappeasable grounds and for the most outrageously ridiculous reasons the Newtonians can think of.

I explain how gravity forms but I am rejected because they are of the opinion that my work does not meet. One such an article I may use because I said I was going to use the material as an open letter I gladly show. I submitted an article in which I show what the manner is in which gravity conducts movement by means of singularity. I wish to produce as evidence e- mail response I received when I contacted members of the Physics Academic establishment and show my case while also showing the response I received. Readers will find witness of what I accuse science of and their demeaning attitude from physicist wishing to silence me at any cost.

Have you, the person reading this, ever thought how it is possible to see that much information that you see at night when looking at the sky. Ever thought about how you are able to see when you see everything in the night sky and how that much light information can fit into such a small space as your eye? Have you ever sat back and think what the amount of information it is that you see when you see the entirety of the Universe when looking at the Universe at night and what the size is of everything of that which you are able to see?

The one star you see seems to be a near visible dot in the picture while the dot might be hundreds or might even be many thousands of times the size of the sun…and we think of the sun as big. The dot is then that much bigger than the sun because the star we think we see could be a galactica hundreds of times the size of our Milky Way galactica but that shows as in the sky as one little dot and yet that entire structure as big as it is, does also fit into our eye socket.

I WISH TO DEFINE THE CATOGORISING I USE AS PART OF THE BOOK.
I have the utmost admiration for Scientists and I shall never dream of placing me in the same category as academics mainly because of their intellect and achievements. They pushed their corrupt conspiracy of a hoax they present as science and which they further by brutal brainwashing through 300 years of never getting detected and that in itself is an achievement unheard of in human history. That achievement is most brilliant and no religion of magical mysteries in the past could ever match the Newtonians. Every time I go against Mainstream Science which is another name for upholding Newtonian blindness I am told I do not seem to have the intellect or mental capacity to *"understand Newton's classical mechanics"* and then because of my limited vision on physics I should know my place and retire to a dark corner where I would then silently and quietly vanish from earthly records. They forever tell me there are two positions on earth: those with the mental capacity *to understand Newton* and then there are those in my sector *that is mindless to the point of not understanding Newton.* In that sense there are two classes, the clever ones that *understand Newton* and then me, the mindless that just cant *understand Newton.*

Can you think of any religion that spread its entire belief on self-promotion without ever proving one validation? Can you think of some gospel preacher going around telling facts as if submitting the utter proof and the entire base is not even valid as a semi believable rumour? ...And when there are Doubting Thomas's you turn the untruth of the system to reflect on their stupidity? They are brilliant! They are the true masters of deception and then they think Devil worshiping is deception. I am going to put a statement and to prove how brainwashed all person are, I think not one will catch on to what I am saying. All things fall equal notwithstanding size. I giant battle tank will fall at the same pace as an empty drum falling next to it. We see this on television almost everyday in all sorts of advertising. Therefore when things fall weight and size does not carry any differentiation as to the falling process. **Yet it is *"mass"* that pulls the falling object to create the gravity or the fall.** Everything falls the same in free fall. Then you explain to me how the F$&@ck can *"mass"* have anything to do with what causes the falling of objects. The fall is neutral in size and in weight but linear movement brings about the stopping of the fall and launch flying? Those physicists don't prove anything because they brainwash everything into submitting.

To substantiate this segregation I use some referring to place distinction between the highly schooled super trained academics that spent most if not all of their lives in preparing to further their minds, filling it with the same void they fill the Universe and calling it "nothing". When I asked where is more nothing: Between Pluto and the sun because Pluto is the furthers from the sun holding the most "nothing" between it and the sun or in the centre of the sun because there is nothing standing between the sun's centre and the centre of the sun, I was discredited as incoherent and irrational. I tip the opposite of the scale as I spent little time repeating the brainwashing they subdue every student with to believe in the norms taught as the official policy in learning and education I have to be on the "other end". I don't believe their crap and tell it as I see it and therefore I am dumber than a pig, or that is their opinion. From where I stand and admire those in science, I can only see intellect as they fooled every person on earth for centuries non-stop: and moreover that achievement is presented as the academic's common denominator. If that is the common denominator used on the one side, fooling everyone by using unsubstantiated rumours and gossip and putting that as the joining factor, then on the other side, which has to be *"my side"* must then be the class of stupidity. To those forming the brilliance in science and their class such a remark would be an insult but to me (and therefore my class) it rings truth and that makes it not an insult but a norm we should except and learn to live with. I would rather be stupid and not *understand Newton* than be **Brainy** and believe I *understand Newton...* how stupid must I be before I would be able to *understand Newton.*

It is rather a pity that while the SUPER CLASS will never say it to our faces; the SUPER CLASS is strongly of such opinion that we on the other side of the Universe have no minds to think in any way, and it is therefore our duty as much as it is our absolute privilege to except what the SUPER–EDUCATED, the ones occupying the informed side of the Universe inform us to what we should accept and the SUPER–EDUCATED live by that idea. As I said I have to live with it too and if I am the ill literate, then the SUPER CLASS must be the SUPER–EDUCATED; where I am the class amounting to stupidity the SUPER CLASS must be the Brainy Bunch. It all comes from the fact that there is such a huge differentiation between us. Those that *understand Newton* is therefore Superior and I, that don't *understand Newton,* are of the lesser blessed. To distinctly point to grouping or class or whatever the readers wish to consider the division there are between the SUPER–EDUCATED and me I refer to the SUPER–EDUCATED side of the Universe by the names I use above. Further more when I refer to mistakes that I do prove to be mistakes in the book as we go along I refer to it as Xepted mistakes to clear another distinction of necessity. In short I don't **_understand Newton_** and therefore I am stupid and they **_understand Newton_** and therefore they are brilliant and what I present must hold the categories in such class divisions.

This is what we use to see the entirety that we call everything or the Universe. How does the photons manage to convey one complete picture coming from as far apart and as wide an area as it does?

I repeat this again to stress the thoughts with which I started but this is how I start: My introduction as well as introducing the readers to general cosmology has to be in a very brief and compressed manner but first, I have to give the emphatic warning to all prospective contemplating readers. Should any person have any doubt about my statements do go to the Internet and confirm what I say about science is true. Trying just to convey my message after compiling my ideas kept me busy for the past going on to twelve years on full time basis whereby I was trying to introduce my findings to many academics without finding much joy from my efforts. This past eleven years plus saw me go without any income as I tried to get my theorem recognised as well as get my warning noted. I just have to get a financial return on my efforts.

Going without a steady income left me almost destitute and in order to find a manner to get my theory across to the attention of influential readers, I decided to publish the theses of 8 books electronically as to try and get around the stranglehold of Newtonian bias controlling science at present worldwide. I decided to publish electronically which those in power do not control. However to get people to believe me is to change science that everyone believes as culture. With my first language not English and the books not linguistically checked by an expert there are bound to be language errors that readers will notice. In the past I tried to check my work myself but after checking say one hundred and fifty pages for language corrections, then after days of toiling instead of having corrected work I ended having four hundred pages of newly written information which is still not linguistically corrected but holds a lot more information.

The language and spelling errors compiled instead of reduced. This is because my priorities lie elsewhere. I aim to spend money on correcting the work as far as language goes, as I receive money in the selling of my theses and in the hope that I will receive money. I will have all my work including the one you are reading edited professionally and corrected as I find money to do so...But first I have to get the public aware of the problem to get the academics to appreciate the problem. In everyone's mind science is more perfect than any form of religion. In the event of any readers who may have questions concerning more facts as it is presented in this book; please feel free to contact me, PEET SCHUTTE. All information divulged came about through independent self-study during the past thirty-two years or so. I have to warn the readers that the topics are showing a very new approach with no quick answers abstaining from proof or holding just a few lines and the information is new in nature but does require much concentration.

I challenge anyone who disputes any claim I make to prove me wrong by proving me wrong and not merely suggesting claims in that direction. I furthermore challenge any person to prove the solar system or indeed stars form or operate in the manner of applying mass as the Newtonians teach. I challenge every person to prove that the Titius Bode law is not the way the solar system forms. Whatever they dispute or however they ignore me I am the only one since time began that was able to explain as much as I also do prove how the solar system forms as I explain and prove how the Titius Bode forms. But the Titius Bode can only form by implicating the Roche limit, the Coanda effect and the Lagrangian points and notwithstanding the importance of my uncovering of science principles that were never yet understood before science and Scientists ignore me and give me the cold shoulder. I can explain gravity for the first time ever and in science no one shows interest because if they give me any credit for being correct then they have to admit that all the information they approved so sincerely was falsified facts implemented since the time of Newton and think of all the rotten egg they then will have dripping from their faces.

My work changes everything anybody ever understood about God's Creation and when going deep enough I don't only prove HOW God created the Universe but I prove mathematically that there is a God and only because of this God being the Creator could the Universe have been Created by His thoughts, but that is going deep into my theory. In another book I show that the Universe started exactly and precisely as the Bible says it did and if you change one word the mathematics I use does not add up. This I accomplished because I mathematically found the manner in which the Universe formed before material formed. I can explain how the four pillars hold the Universe in place and how the four pillars form in the process they form a Universe. The Big Bang only came about when light formed the Universe and there became departmental fraction between everything that spins faster then the speed of light such as neutrons and protons and that which spins as fast as the speed of light which is electrons and photons and then all other space forms gas as it moves way slower that the speed of light. This is when the Bible sates the Creator ordered, "Let there be light" but there already was a Universe in place by that time.

In the name of the truth in future science, I hope you can help in supplying a name of an investor that would pay for the promotion of the publishing that will enable the book to sell. However keep in mind that this book or any of my other books for that matter is not immediately going change science because science takes about 50 to 80 years to change if history is repeated. Not one person in science will rush to my aid and admit science has been a fraud this past three hundred years. What this book intends to do is to get people in some sectors to begin to think and to question those who hold positions so high that they are deemed to be beyond questioning.

Everybody think of the four cosmic pillars as working that far away but the four cosmic pillars control our every day life by controlling our movement as it regulates the gravity effecting our every movement.

CURCULAR ATMOSPHERIC DISPLACEMENT $\Pi^2 \backslash \Pi$

Atmosphere

$\Pi \backslash \Pi^0$
LINEAR DISPLACEMENT

$\Pi^3 \backslash \Pi^2$
AIRCRAFT
TOTAL DISPLACE-
MENT

Everything forming gravity applies the Titius Bode law. When an object stands on earth holding "mass" the object serves a position as a unit of the earth while it moves with the earth. This is part of the four cosmic principles forming a unit that forms gravity I named on the previous page.

The reason why an atmosphere forms is the Lagrangian points forming atmospheric boundaries with density limitations applying.

When the objects starts to move in a horizontal fashion that exceeds the gravity going in a vertical fashion the Coanda Effect bring about that the object then forms a part of the atmosphere where it releases from the earth and becomes part of the atmosphere. We call this release flying.

BETWEEN MACH ONE AND THREE Π^3 / Π^2
ATMOSPHERIC CIRCULER DISPLACEMENT $\Pi^2 \backslash \Pi^0$
ADAPTED   CIRCULER DISPLACEMENT
$\Pi^2 \backslash \Pi^0$

Atmosphere

$\Pi \backslash \Pi^0$

LINEAR DISPLACEMENT

Then with much more movement applying the aircraft breaks the sound barrier. This is actually where the movement exceeds the Roche limit and in that movement creates that the moving object forms an independent atmosphere.

This is when as the concentrates so much

Above the Roche limit of $(\Pi/2)^2$.
CIRCULER DISPLACEMENT
$R^2 \backslash T$
ADAPTED

Atmosphere

R/T LINEAR
DISPLACE-
MENT

Roche limit the aircraft by moving independence it captures a cloud of vapour by concentrating the density of the atmosphere that the aircraft forms as a unit to become a unit within the earth's unit but an identifiable part that is no longer part of the earth's atmosphere. That is why at that point it carries its own rules on sound outside the borders of sound on earth. Very briefly this is the information about physics that the four cosmic principles or laws provide us on gravity. Gravity is immensely more than simply "mass" pulling "mass" that then bring a number of complex formulas and equations about.

Every speed we travel in places us while we travel in a different size bubble and every size bubble has its own limits preventing us from going onto the next bubble or back to the previous bubble. However I hav e books explaining this in detail but this book is meant to introduce my ideas and not explain or prove my ideas.

This is nature and you find this in nature and nothing about Newton forms part of nature.

This is the Titius Bode law: The Titius Bode law proves that mass has no place in science. See in the picture how random mass is and with such randomness, how can mass place planets in the positions they hold? By my effort to solve the mystery of the Titius Bode Law, I prove that gravity forms not by mass but gravity forms by π forming in movement π². Solving the Titius Bode Law and proving from that how gravity works opens up a new view on the cosmos.

This is **The Roche limit:** The Roche limit has been around for centuries and with all the mathematical splendour available to apply in order to fathom concepts behind this phenomenon, still with all the computing ability of a machine all those physicists with all the mathematical superiority could not touch any understanding about the concept forming the background. Yet when using the truth about gravity in physics the answer is simple; it is that gravity is Π.

This is the Lagrangian points :The Lagrangian points have been known to science for centuries and with all the mathematical splendour available not one calculation could ever explain why this event is taking place. The satellites form precise locations positioned around the major planet and never comes closer while remaining in their positions.

This is the Coanda effect : The Coanda effect has powered turbine engines and aeroplanes in flight for almost a century and with all the mathematical splendour available to design the most terrific aircraft, not one engineer could mathematically compute one fact to show understanding why this takes place. How sad it is that those claiming of much superior intellect in physics remain just no more than having computing power.

The understanding is not complex. I have to warn the readers that the topics are showing a very new approach with no quick answers. Understanding is in the proof and that does not come by reading just a few lines and then forming conclusions. The information is new but not hard to grasp. I did not put these phenomena in place and these phenomena nullify Newton's correctness and the proof I bring goes beyond any doubt. I prove the Titius Bode law. Go to the internet and see how science doubt the Titius Bode Law and the correctness thereof while to solve the problem you add 3 plus 4 to get 7. That is if you want to find a solution. I have published the Titius Bode Law in four already published books but in this one I go deeper than the four already published. In each of the books I present I disclose how the Titius Bode Law forms gravity.

These four laws also constitute to gravity we experience on earth and this is how…

I wish to give a small example how bad science fools the public and the public allows science to get away. This is one example and I can and do mention many, many more in all the books I have on the market. You've heard the story that a car lifts into the air because underneath the car lifts the car into the air. It is air flowing increasingly underneath the car that supposedly pushes the car up and therefore the car lifts into the air. What a cock-up story this is. Think about it when using a little bit of intelligence behind the idea. It is so easy to see this entire concept is filled with fraud, Newtonian fraud.

There s forever more air flowing over the car than what there could ever be coming underneath the car.

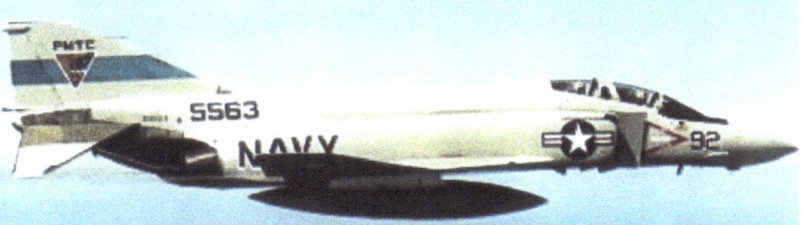

The car is pushed down by wind many times more at any point than air could be underneath the car lifting it.

This argument gets even more valid when the car goes faster. The air flowing over the car that pushes the car down is so much more than the little bit that can flow underneath the car. So how can the air underneath be able to lift the car against the pushing

the car down with so much thrust. Just a small thought and all that Newtonian wisdom flies out the window showing how much crap science feeds the public. This argument I explain later at some point in this book in more detail. I mention this to validate the reason why you should purchase this book.

I have studied gravity and more in particular how nature applies gravity. Nature works by applying the four cosmic pillars or cosmic laws I have mentioned. Gravity works on density that is between filled space that moves and unfilled space, which the solid moves through.

The faster an object moves the more the ratio becomes between the space it holds and claims per time unit and the space that it moves through per same time unit. This forms a variable density that changes as the speed increases. If you put the ratio on the aircraft, then the structure of the aircraft reduces because the ratio then increases in favour of the increase in space per time unit the

$$7(3\Pi^2) \text{ X } 2\Pi^0 \Leftrightarrow 7(3\Pi^2) \text{ X } 4\Pi^0 = 829$$

$$7(3\Pi^2) \text{ X } 2\Pi^0 \Leftrightarrow 7(3\Pi^2) \text{ X } \Pi^2 / 2 = 1022.79$$

object moves through. When the relevancy is placed on the air the object moves in then the object increase in size as the speed increases. The distance the object moves per time unit becomes more increases and the ratio between unoccupied space and occupied space increases which increases. To leave the earth gravity or atmosphere a relevancy of movement between gravity moving space towards the earth and a body moving against this requires $4\Pi^2(7°(3\Pi^2)(\Pi^2/2)/6^2\text{x}10) = 11.365$ **km / sec.**

$4\Pi^2$ **Earth's atmosphere**

The relevancy required too leave earth's gravity is $4\Pi^2(7°(3\Pi^2)(\Pi^2/2)/6^2\text{x}10) = 11.365$ **km / s.**

$(\Pi^2/2)$
Roche limit

$(7°(3\Pi^2)$
Earth rotation

Divided by $6^2\text{x}10)$ **earth's curvature**

The space the object holds becomes larger but in relation the space the object holds becomes less in relevancy to the space which it moves within. In other words if one concentrates on the solid then the solid shrinks and when the concentration falls on the air then the object increases in size. Therefore there is a point that the moving object does not fit into the space it holds on earth and must get airborne to claim more space in ratio of the size it holds. More movement require more space to fill per time unit.

Speed is relative to space filled by movement by time in the instant.

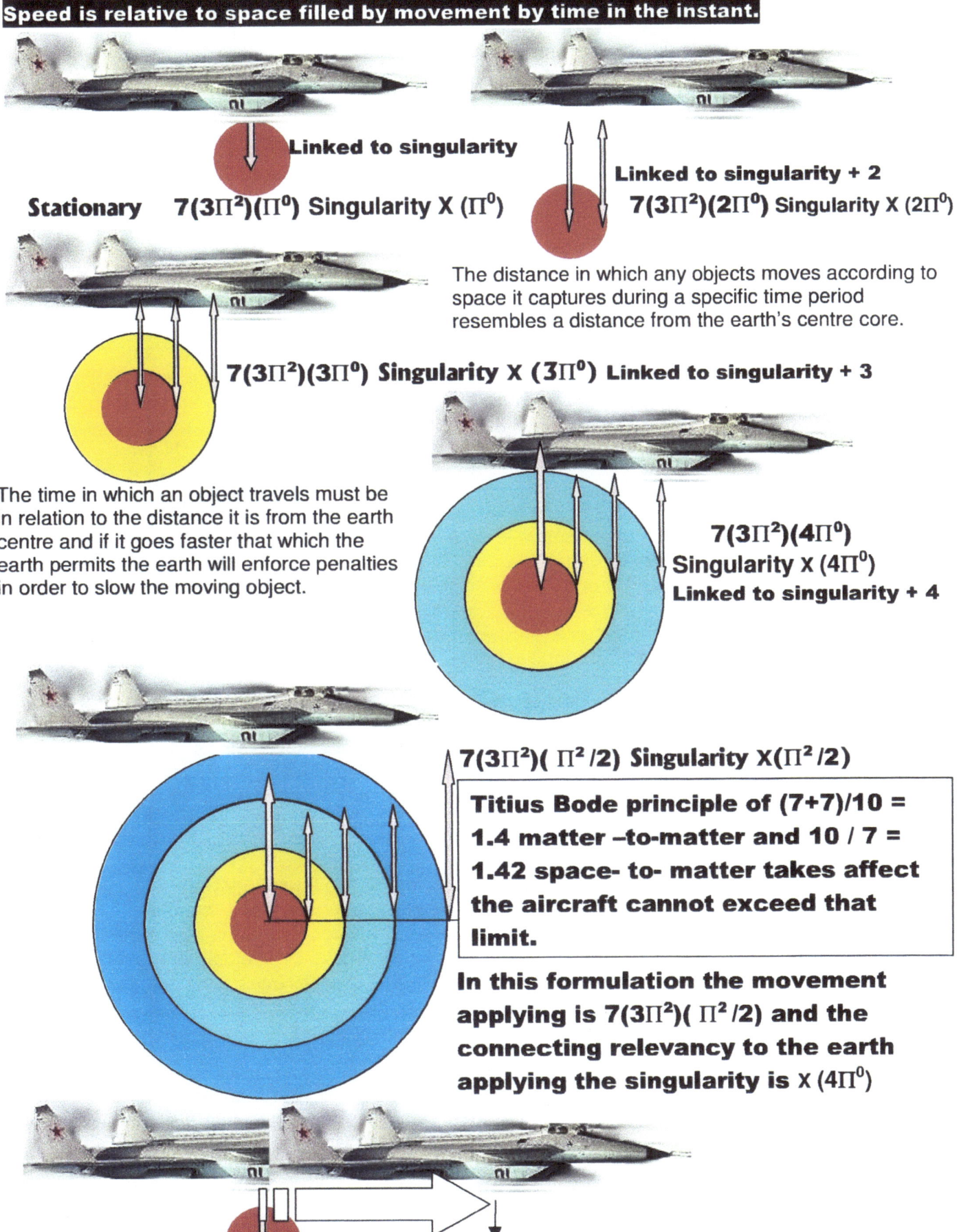

Linked to singularity

Stationary $7(3\Pi^2)(\Pi^0)$ **Singularity X** (Π^0)

Linked to singularity + 2

$7(3\Pi^2)(2\Pi^0)$ **Singularity X** $(2\Pi^0)$

The distance in which any objects moves according to space it captures during a specific time period resembles a distance from the earth's centre core.

$7(3\Pi^2)(3\Pi^0)$ **Singularity X** $(3\Pi^0)$ **Linked to singularity + 3**

The time in which an object travels must be in relation to the distance it is from the earth centre and if it goes faster that which the earth permits the earth will enforce penalties in order to slow the moving object.

$7(3\Pi^2)(4\Pi^0)$
Singularity X $(4\Pi^0)$
Linked to singularity + 4

$7(3\Pi^2)(\Pi^2/2)$ **Singularity** $X(\Pi^2/2)$

Titius Bode principle of (7+7)/10 = 1.4 matter –to-matter and 10 / 7 = 1.42 space- to- matter takes affect the aircraft cannot exceed that limit.

In this formulation the movement applying is $7(3\Pi^2)(\Pi^2/2)$ **and the connecting relevancy to the earth applying the singularity is** $X(4\Pi^0)$

$7(3\Pi^2)$ (Π^0)

Singularity X (Π) The aircraft ends its propelling potential and thrust require heat artificially generated to further any increase in velocity

Should you wish to learn more about this there are books that concentrates on this aspect and in which I explain this in much detail. My introduction as well as introducing the readers to general cosmology has to be in a very brief and in a compressed manner but first, I have to give the emphatic warning to all prospective contemplating readers. I didn't make up those ideas I attribute to mainstream science as I went along but it is Newtonian science that cling onto the black magic Newton believed in. I don't wish to explain what science regard as the truth because I have too much of my own that is correct to introduce and to explain.

I wish to make one fact very clear. I base my work on formulating the working process of four cosmic principles in Nature. These are:
1) The Coanda effect
2) The Titius Bode law
3) The Roche limit
4) The Lagrangian points.

I did not discover these phenomena because science knows about these phenomena for a very long time and in some cases even for hundreds of years. Science knows they apply and where they apply. When science discovered or allocated missing planets they used the law applying such as the Titius Bode law from which they deducted positions that they knew in that circle according to the planetary layout that the law predicts there had to be a planet according to the law. Science did not apply Newton's formula to discover and locate planets but they applied these phenomena and especially applied the law of planetary allocation to discover the precise location the planets discovered after Galileo.

Every one in science knows these phenomena is there and is in place and they rule the orbit set-up of the planets. The solar system functions according to them. These four laws on planetary motion that is used by nature at this moment and has been in place since time began are what apply and they dismiss Newton. If you argue with me about Newton being correct you better take your case to God or the solar system because the four cosmic phenomena is working in nature and nothing Newton said is applying in nature. This is a truth and a fact and a foregone conclusion and can never to be in doubt.

Brainwashed as you may be in believing Newton you can't either side with my view or decide on Newton because it is not a case of choosing between Newton and me. I'm out of the picture! It is either telling nature to listen to Newton and change what is in place or read and see what is in place and what is applying in nature all along. The phenomena are what we find to be used in the cosmos while Newton is in the imagination part of the minds of scientists and nowhere else. If you don't believe me and if you wish to discredit me first find out a little more about science. Then deflate your ego as to what you think you know.

Science never mentions these phenomena because science can't use Newton and explain these phenomena or use these phenomena to prove Newton. These four phenomena that the cosmos uses as we speak have been in place ever since the Universe formed. Since science can't explain the phenomena and the phenomena destroy the credibility of Newton science avoids these phenomena as if it brought the plague. You can't choose between Newton and me because I did not put the cosmic phenomena in place. All I did was doing a study since 1977 to formulate how and why these phenomena work and how these phenomena keep the cosmos and the solar system working. I am the first in history to show why they work.

We are all been brainwashed for centuries to believe Newton. Should your brainwashing kick in and you have an axe to grind with me about what I say, then first prove these phenomena are not in the cosmos and are not applying to form the laws that the cosmos put in place as gravity. I only found out how they work and why they work and I did not make the phenomena work. All those clever stooges that have so much to say even before you read first learn what is in place before getting so opinionated.

I am not fighting science or the credibility of Newton or what might be true or not true but I am fighting centuries of brainwashing and I have to dismiss the brainwashing and the systematic mind control that those teaching science inflicted on us all. My fight is not about what is true or not true but what is accepted as culture and which was not even once been proven by science. Please do read on to investigate before becoming self-opinionated.

Revealing the book that aims to become the most ardent whistleblower on

a Conspiracy

In Science In

Progress

There are two identical other letters published at, in e-book with the same motive as this in which the one supplies **more information** than the other one does and other option is a **lesser one as it is shorter** than what the more comprehensive addition is,

To whom it may concern:

I am P.S.J Schutte, nicknamed Peet. Being a white South African my mother tongue is Afrikaans and my second language is English. My language standard used might not quite meet your Shakespearian expectations and some spelling errors might have slipped through cracks but it is not the art of language I try to promote for it is the sharing with you the knowledge why science never disclose information they prefer not to reveal and never to divulging the entire overall picture about what it is that science knows and don't know. Science perfected the art of hiding the truth. I discovered the four building blocks nature uses to build the Universe. Was I saluted for my efforts...no, because my findings trash their accepted views that are based on Newton! I now turn to the public for help in my fight against those with absolute power and they are accountable to no one or answerable for any veracity of what they dream up as truth.

With my first language not English and the books not linguistically checked by an expert there are bound to be language errors that readers will notice. In the past I tried to check my work myself but after checking say one hundred and fifty pages for language corrections, then after days of toiling instead of having corrected work I ended having four hundred pages of newly written information which is still not linguistically corrected but holds a lot more information. The language and spelling errors compiled instead of reduced. This is because my priorities lie elsewhere. I aim to spend money on correcting the work as far as language goes, as I receive money in the selling of my theses and in the hope that I will receive money. I will have all my work including the one you are reading edited professionally and corrected as I find money to do so...But first I have to get the public aware of the problem to get the academics to appreciate the problem...but after you read this letter you'll know why they have an attitude.

What I am about to explain is not how the cosmos started because that is immeasurably more complex than the depth I am going into with this conversation. My intensions are to show why singularity is so important, why seven is connected to the birth of the Universe and how ten became the liquid of space.

Looking at the night sky you see many flickering dots and spots and light coming from afar. The one star you see seems to be a near visible dot in the picture while the dot might be hundreds or might even be many thousands of times the size of the sun...and we think of the sun as big. The dot is then that much bigger than the sun because the star we think we see could be a galactica hundreds of times the size of our Milky Way galactica but that shows in the sky as one little dot and yet that entire structure as big as it is many times our Milky Way, fit into our eye socket. But that is not all...there are trillions of such light images and they all fit into one eye socket. What we see is immeasurable and yet we see it effortlessly in the space our eye holds...how can that be? It is because we see dots and spots that form a Universe.

I have asked this many times before but I ask again: Have you, the person reading this, ever thought how it is possible to see that much information that you see at night when looking at the sky with only your eyes? Ever thought about how you are able to see when you see everything in the night sky and how that much light information can fit into such a small space as your eye? Have you ever sat back and think what vastness in information it is that you see when you see the entirety of the Universe when looking at the Universe at night and what the size is of everything of that which you are able to see?

But that is not all...there are trillions of such light images and they all fit into one eye socket. What we see is immeasurable and yet we see it effortlessly in the space our eye holds...how can that be? How is it possible to fit what we see into the space of our eyes we have? Think how much is the entire information that is visible at night and think about how all of that fit into the space your eye holds? This is what the Universe is and that is what the Universe represents and when I put that into a mathematical equation I wish to see those persons designing space whirls to calculate the volumetric space reduction with physics.

Consider how big is what is visible and put that space into the size of what your eye can hold and ask your mathematically educated Professor in physics to find some ratio between what you observe and the size of your eye. The ratio is astonishing, but more-over what is truly astonishing is the arrogance of man to think of his position, as being important while the space man holds is beyond any comparison in ratio to everything we see in the Universe we see. Think how small we are when we are able to see the entirety out there! Even if there was other life out there, what is the worth of it in comparison to what there is that we see? Tell your physics professor not to reinvent the Universe will his brilliant mathematics, but just too mathematically formulate how all the light forming a Universe, the visible Universe can fit into my eye.

How is it possible to fit what we see into the space of our eyes we have? Think how much is the entire information that is visible at night and think about how all of that fit into the space your eye holds? We tell the Universe is must contract or expand and that it applies mass. In this idea about how you are able to see the entire Universe you will find all the answers to the questions about how physics use time to employ gravity. Mass and anything Newton ever said has no implications on the explaining. If you question this remark then use anything Newton ever said that can explain how it is possible for a human eye to take in all the information that the Universe can provide in a glance at the Universe. I can and I do explain that but only after discarding Newton's ideas. I mathematically prove how it is possible to witness all the information of the entire Universe presented in one electron contacting a nerve in your eye.

No less telling is the next thought I present you with. Why would all the light coming from all over and everywhere meet you in the precise location you hold? The light left its place of origin and travelled for in some cases 12 billion years or more at the speed of light no less to meet you in the position you are at the location you are and all the light coming from everywhere possible comes directly to you. Makes you feel important does it not? What a thought this is for the further boosting of the ego that man cultivates!

It is not what science declares that is important but it is always what scientists don't declare that holds prominence and more so the reason why science keeps a silence about the information they do not disclose. It is never about what they say but it is why they don't say other things they keep quiet about. You will read how they never disclose the entire truth because science is about promoting one-sided and selectively opinionated information forming fraud no less. I have been per suiting a new cosmic theory that I partly present in an eight part theses, of which the investigating research began in 1977. In 1999 I compiled my theory and searched for a publisher. I use Kepler's formula but not as Newton did when he raped the living daylights out of it because he had no idea what it was about. By using Kepler's tables correctly I show where to locate the very first spot that forms the point where the Universe started. The question about the Universe is how can whatever is in view, come stored as a parcel in an electron, and tell the entire story about the entirety out there locking all that data into the space of a photon I use to see things. That is physics and however you may try, not one person searched for the ability to calculate that part, although the cosmos gave Kepler the formula.

I decided to place to letter as introductions to my quest in getting a promoter or financial backer. To get my message across is not very easy. Insight and comprehension is eyes that allow a person to see what others with less insight and comprehension are unable to see. The more insight and comprehension a person has, the more detail can a person observe but also the less insight and comprehension such a person has and then to that person the less information exist in the eyes of such a person. Those persons which science frown upon as having little insight and comprehension are deemed by science as to be brainless fools and I am going to prove this with much clarity and they are according to science to be closer being unsighted where with the absence of insight and comprehension such a person is bordering the limit of sightedness and what is being completely blind. This is how science see us all that they think of as inferior but in my work I have to cater for all persons from those that have immediate and clear vision to those requiring more time to observe and to take in and are satisfied with less information because they accept more easily than others do. However, I believe science must be open and discussed by everyone and for that reason I have to cater for many different levels of understanding information.

This has no implication on intellect because in my book I show how quickly some can jump to conclusions while they are unable to read and understand two paragraphs but still feel the need to have an opinion notwithstanding how uninformed and unsuitable such views might seem to be in the manner it may come across. Some I accuse are members of my immediate family. If you are unable to understand what I write you can see less than what I can explain and then it is because you cope with less information because of lesser-developed insight but that does not make you intellectually inferior. To those that are satisfied with less information I leave you that option and you can read the shorter version of the letters on offer.

The letters are precisely the same but one states more facts and inform more about more topics than does shorter versions. What makes any person seem to be intellectually inferior is to find that person is opinionated by a subject the person has no idea about and then when knowing nothing about what is said then still feel informed and draw irrelevant observations while not being able to understand a word that was said. To those I say, read again before your stupidity overwhelms you and your poor judgement runs off with your better judgment and you come up with remarks clearly showing you are out of your depth.

Science is doing to everybody the same injustice and therefore I wish to inform as many persons on as many levels of understanding as I can manage. To get persons to read and understand what I inform about science I wrote two different introduction letters, each with a unique and different levels of informing while delivering the very same information. I started with the first letter just to inform as to interest possible potential investors. As I wrote on and started to explain something I knew was new to a reader, then as I got to a point where I knew the information must come across as very different to what is the norm that is accepted as science and form the cultural basis of science then to soften the surprise I touched another aspect every time because understanding more about the topic was a critical aspect of information needed to fully understand what I tried to say. This made what I was conveying very bulky and it would therefore seem to some as being top heavy. My wife is very quick to show me that I over inform those like her that would rather prefer to be under informed. The facts can accumulate to a point where some persons get the feeling they are swept away and are drowned by the number of facts I share. There is no room to present even the least bit of proof in any form possible in the space given to this article. With the limited space available to publish information in a journal by way of a small article such as this and having so much information at a premium I decided to release some vital information and the required proof about my claims in comprehensive works that can be obtained elsewhere.

As the letter evolved I found some other or even more information that I also have to share before what I shared in the context that I was explaining was fully understood and accepted. But before finishing the part that I was explaining at first I realised what I was explaining at that point would be better understood by any reader and I found myself deciding I have to bring in more information to leave the reader better informed about what I was conveying...until I came to another aspect and at that point I realised this was even more important to include than what I first included just before I explained the next bit. I came to a point where I had to decide to include information I started with and then to include information I included just before the information that I started with and by the time I came upon the point I wished to explain I realised the first was even more important than was the what I explained. In the end what you were reading as the conclusion to the letter I was on my way writing another too complicated theses for any one to enjoy. What you are about to read is very much as new as anything new you will ever come across. Therefore I give you the choice about how much introducing you feel comfortable with.

You are the first to read what you are going to read for the first time. Now I give you two letters providing three options to approach the information I would like you to read. There is a choice about what you decide to read (for the very first time) that informs what is necessary to know and from that to form an opinion and the other is to leave you well informed but reading this option takes much more time and effort and it requires more concentration to enjoy what I convey. I now decided to leave it up to your decision which version you feel comfortable and with which you would like to start.

Whatever your studies were or in which direction your field of studies developed your interest or to what ever level of involvement you rose in albeit physics, cosmology, architecture, engineering or theology or whether you did no studies at all and notwithstanding the level or to the degree your studies came to evolve too, it will be of very little worth in help as you now are about to embark on ideas putting something forward that is completely new to science. Any such knowledge that you may have gained in the past about science founded on the Newtonian direction of thought would come to zero, as the work you are about to embark on is as new as anything ever could be...and in that sense it would be better to decide the level in on which you wish to engage and to first familiarise yourself with information. First, before starting to read, read what I present in these letters I present as to get confident with this new work.

I show what is wrong and still there are those that accuse me of being wrong just because the "honourable" scientists who among all those are also including physicist just could not be wrong according to the brainwashed masses and other zombies studying physics. Yes and saying brainwashing is no cheap talk or shouting wolf just to draw your attention. You are going to see how they involve all persons on earth including me, you, your parents and if you have your children then them more so in mind bending thought manipulation and brainwashing and again I say this is no cheap talking or a drunkard talking. However the understanding of the facts I present and to the degree of acceptance you may ask for is in a degree or level of informing and carrying across facts representing the truth.

You can start at any one of the two options and read that which you decide on. I leave you this option because I have no idea what people may deem as to be complicated and what people may find easy reading...to me it is all the same because I have been living with all this information you are about to read and then about ten times more for most of my adult life. However, there might be those that think what you read is simple but for those I say: believe me this letter does not even form the idea that will result in an introduction considering what information the rest of my work presents. I have endured criticism by persons that clearly have the science insight of a skunk and the understanding about science of a porcupine and they would not be able to read one paragraph of my Theses I named *Matter's Time In Space: The Theses.* If anyone wishes to criticize me you are welcome but present facts and not an idiot's slurring. Why this response: It is your attitude that helps to keep cheaters in office conspiring to keep the human minds unaware of a very important truth they successfully hide. No matter how informed you think you are, if you aren't able to explain how the cosmos started, not from the simplistic rhetoric Big Bang, but from where the first dot became the second dot and then the third dot, and why the first dot became the second dot and then the third dot then just remember I can and I did. In some of my books I explain how this happened. I will leave you a clue; it is because 1 and 2 is 3 and 2 + 2 = 4 while 2 x 2 = 4. This configuration in why mathematics is a line using numbers might seem simple, but that is how it all started and it's more complicated than anything you think of as being science.

My work changes everything anybody ever understood about God's Creation and when going deep enough I don't only prove HOW God created the Universe but I prove mathematically that there is a God and only because of this God being the Creator could the Universe have been Created by His thoughts, but that is going deep into my concept. In another book I show that the Universe started exactly and precisely as the Bible says it did and if you change one word the mathematics I use does not add up. This I accomplished because I mathematically found the manner in which the Universe formed before material formed. I can explain how the four pillars hold the Universe in place and how the four pillars form in the process they form a Universe. The Big Bang only came about when light formed the Universe and there became a partition between everything that spins faster than the speed of light in atoms holding neutrons and protons and we have that which spins as fast as the speed of light which is electrons and photons and then all other space forms gas as it moves way slower that the speed of light. This is why the Bible states the Creator ordered, "Let there be light" but there already was a Universe in place by that time.

In the name of the truth in future science, I hope there are those with the means and the will that can help or supply a name of an investor that would pay for the promotion of the publishing that will enable the book to sell. However, keep in mind that this book or any of my other books for that matter is not immediately going to change science because science takes about 50 to 80 years to change if history is repeated. Not one person in science will rush to my aid and admit science has been a fraud this past three hundred years. What this book intends to do is to get people in some sectors to begin to think and to question those who hold positions so high that they are deemed to be beyond questioning.

Previously I did a project to get a feeling of what is going on in the publishing field and to test if what I write about is feasible and is viable in any commercial environment. I have work published at Lulu but also I was informed that I have eighty percent of all the work published on Lulu.com and that tells me no one interested in buying works about science is going to Lul.com for information. Then I published two none descript books on a website that entertains by offering books free of charge. I don't belittle the website but I just tell the truth. The truth is the higher a persons is developed the more information costs.

Normally through the centuries it takes about 80 to 100 years for those in science to conform to a new idea and then confirm the new idea. Going back in history we find that all those locked onto old and outdated conceptions had to die out and be replaced by much younger generations, more advanced with less to lose before a new idea was confirmed and the old idea became absolute. I thought in this modern era this would not be the case but Academics have a stronger grip on publication in modern times than what the Pope had in the time of Galileo. At least the Pope allowed Galileo a chance to publish his thoughts but that is more than what modern liberal minded Newtonians would allow. Their stranglehold is vicious and murderous when it hides the truth and in that I need the help of a publisher that can confront those who need confronting to break this iron grip and show what mockery science is at this point.

However, I just can't see being in this age of the Internet and electronic publishing that I have to wait to become noticed eighty years after my death! My case is solid and what I bring to the table is beyond denial but because they are formidably strong I stand no chance man-alone. Even in these few pages anyone with some form of perception about what applies and what does not apply must see what I convey. I need some publisher with a backbone that can help me take on the meaningless science and introduce the truth for the first time ever. Those formulating science are quick to throw stones at the Pope but they are exactly as guilty. When someone does not agree with their views they go out to destroy you.

With all the amazing achievements accounted for and when recognising all that science changed our way of living on the earth and what was achieved by scientists developing this super mentality and in that also giving science all the admiration dually admitted, notwithstanding I am about to dump on you the biggest conspiracy that has ever been presented and that was ever undertaken by any group of persons in the entire human race. Think of anything you might think is big or outlandish by nature and that dwarfs in comparison to what I am about to reveal. It is so large that there is nothing in the past history of man with which one could compare it to. It involves every aspect of the life of every human being and this shadow in our midst covers the darkest secret that was ever hidden from intellectual human view. It is perpetrated by those we absolutely unconditionally trust in all aspects, It touched on every individual walking the surface of the earth and that excludes no person of any status albeit it an infant or someone in old age.

Newton introduced mass as a brilliant concept. Mass is that which makes you stick on the Earth and gravity is that which pushes you onto the Earth and weight which is the same thing as mass is what you get when gravity and not mass keeps you glued standing still on the Earth. Let's forget about the crooked manipulation of the truth and stick to what really apply. When a body can't move further down to the centre of the Earth by using gravity, the body comes into mass. There can't be insufficient mass as they protest to apply in the Critical density debacle. If at the Big Bang there was not sufficient mass to destroy the radius and prevent the expanding from coming about, then the expanding won the match and there can be no contracting Universe as Newton had us to believe. If the Universe started a journey of parting objects no amount of dark matter that might lurk in the night sky and is at this moment hiding from detection will produce the gravity required to stop the expanding from continuing. At the start the expanding became evident and as the radii grows the inclination will suspend in influence as a factor. If there was insufficient mass at the start in order to tilt the balance in favour of the reducing factor, no amount of mass can ever accomplish such a goal afterwards. Then Newton's surmising was one of corruption making that which all physics are based on foolish thoughts and corrupted proof.

I seem to be the first person in generations that ask questions about Newton's work. Questions I now ask is asked for the first time ever, well ever since the time Newton introduced gravity, before the emphasis fell on proof rather than merely what a person with reputation suggested. I now am able to show how gravity forms by forming a circle using Π because I have located the centre of the Universe. But by my effort in finding the location I disrupted everything Academics in physics hold holy and for that I am most unwanted in the presence of the Academics charged with guarding the ethics of physics. In short, I clash head on with Newtonian dogma and principles forming physics. During my research I discovered abnormalities and inconsistencies about mistakes the Arch fathers in physics must be aware of but are hiding with all their considerable influence and academic power. The road I took in my search for truth concerning physics was never smooth and the resistance I came across coming from the academic sector is almost unbearable. I made no friends but only enemies. If you might be of the opinion that my accusing the greatest intellectual department in the world as being in misconduct and to your view such accusing is outrageous and far-fetched, then be my guest and judge the following with a clear and unbiased mind because when scrutinised with a clear view then the facts cannot fool even an idiot. However, that is just what the physics paternity thinks the rest of us forming the general public at large are. They have the opinion that they can feed us in the public arena any senseless rotten garbage they dish up because they see us as being inferior by thought and mind, equal to animals.

You are going to read about a conspiracy but people think of a conspiracy in many terms. Let us define not by definition but by interpretation what a conspiracy constitutes of. What do you think is a conspiracy? All the conspiracies you know about are known about because someone somewhere makes money by allowing the revealing of that conspiracy. Silencing the conspiracy does not make money but informing a suspicious public loosens the flow of money. If it were a true conspiracy no one would know about the conspiracy because the powerful would make money from not revealing the conspiracy. The revealing of the facts about any conspiracy would be stopped before it leaked because it would kill the flow of money, which the conspirators make by not revealing the conspiracy. So then to stop the leak brings the money. There is always a thousand and one conspiracy theories going around and the one tries to be bolder and more sensational than the next theory flying around. However, the biggest conspiracy is going on in front of every person's eyes and is committed by the most respectable persons in any respectable upstanding society. Even if you had your personal favourite conspiracy theory, try and match it to the one that I have! To gauge the truthfulness then see where the money goes by revealing or hiding of the conspiracy.

A conspiracy is thought to be a gossip story that makes money and by not revealing it or revealing it goes in line with making money or not making money. You can download this information free of charge because I don't make money by revealing the conspiracy. The honest truth is I have another agenda. I reveal the information in this letter so selling the information is not my priority motivating this letter. I want to make money but it is by showing how I can correct the flaws in science, not by hiding it in a conspiracy. People put a conspiracy in the same realms as a gossip story, an old wives tail, which is going about but does not intend to harm and mostly serves as amusement to many. Hearing about a conspiracy tests your intellectual comprehension. It is some quiz that you match your wits against the truth that the conspiracy reveals. It is a funny, but it is not funny until you catch the funny part hiding behind the conspiracy and only when you measure the catch behind the conspiracy are you treated to be amused. If the conspiracy does not touch the person directly then no harm is felt and no harm is intended. Every one holds this view that a conspiracy is on a slightly higher level than gossip. It is a gossip story about someone living in the neighbouring village known only to some people next door but has no direct linking to me or has no threat to the safety of others directly associated to me. Everyone treats a conspiracy as if it is something amusing that holds no threat at all. It is something that goes around as a joke of sorts. However, you require a certain intellect to see how you are duped and as we all think we are intelligent it really is what we are fooled with that gives our intellect away. This is a picture of how to hide a conspiracy. Never divulge what you know then no one will find out what you don't know.

When I saw this picture it was terrifying similar to Newtonian science. If in the cosmos there are jumping dolphins then in Newtonian science physicists will force you to believe it is cows that are actually dolphin look-alikes acting as dolphins but in truth are cows. Why would they do that? They will say there are

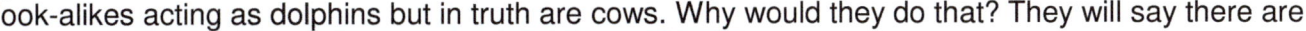

mammals in the sea that feed their calves by producing milk and then leave it at that. Why would they do that? It is because according to what they calculate then their figures show there are mammals. The only mammal they know that feeds calf milk is a cow. That means they are going to force you to believe cows as mammals swim the in the sea. They calculate what they wish to find in the Universe. When they find that that which they find is not what they had in mind after they formulated the equation to get that what they wished for, which is that what should according to their calculations be there and should be in the Universe according to what they calculate, then from these conclusions they will feed you half truths and semi lies as well as fairytales and they don't even know the difference between facts and fiction because they concentrate on the correctness of the mathematics and leave the facts out of the description because they are clueless about what really is in place in the cosmos, therefore they know not better.

However, if a physicist tells a student in class there are mammals swimming in the sea the physicist will present a picture of a mammal in the form of a cow because the physicist knows the cow is a mammal but also knows there are mammals swimming in the sea. Is the physicist wrong? No, there are mammals swimming in the sea and yes a cow is a mammal. Are there cows swimming in the sea…that part he or she will never correct because then everyone will know he or she has no idea what swims in the sea. He or she will leave you believing mammals swim in the sea holding you under the impression that cows are mammals and calves feed on milk in the sea. Only science, Newton and God in that order is never wrong and that is religion.

This acting like God envelops all forms of science. What is the truth? Do we hear the entire truth? The truth is that science only reveals some portion of what they know and ignores what is there that they know they don't know. They only reveal what suits their position and never divulge what they know but what does not compliment their view. When science confronts religion they have the opinion that what is in science is everything there is and there can never be more than what science knows or what science wishes to reveal. If science can't prove the existence of a God then there can't be a God for science can prove everything. Not proving God is not science that is incapable or flawed no, it is God not existing that is the truth. If there was a living God science would prove it but since science can't prove a living God then God Almighty is a rumour thought up by those intellectually slightly less equipped to deal with fear. Science could never be unable and inferior to prove God exists because science is the supreme intellect.

Science knows best!

A Hundred years ago science had this same attitude and today we all see to what extend were they backwards and inferior. Today we laugh at medical practising of a century or two ago and in another century we know the future generations will laugh their heads off when listening to what the informed opinions are of the professionals today. Science has forever veneered their status with this blanket of "they know all". This makes a mockery of the truth because science has no clue why man die or why man age and yet they promise eternal life within the next few decades to come. Ask scientists what is life and they will have an "informed opinion". Well…I wish to bring to mind some of the facts that physics work with when academics as scientists only work with facts. Remember they are the ones boasting that if facts are not proven then facts become fables and those very important academics don't waste time with fables because they only work with facts. Scientists portray the image that they know all there is to know about everything man might ever know and nobody can ever know more than what they currently know.

They tell you what to eat, what to drink, what to think and how to live because they have this image that mentally they are on par with God but with the ability to correct mistakes God made. They are the utmost superiors on all levels of what forms creation. This attitude applies to rocket scientists as much as it applies to medical doctors as much as it applies to lawmen. This comes through in all departments albeit language, art, science, law or whatever you may have. If you smoke cigarettes then you will die young because the medical profession found that it is harmful to you while a hundred years ago people smoked ten times what they smoke today and they lived just as long as they live today. But the medical profession took it on them to act as God on behalf of God and force-feed everyone to do what they say or die.

The truth is the oxygen you can't live without is burning you to death by aging you and without that you die. The most harmful substance you can ever take is oxygen and it is oxygen that you can't live without. If you don't breathe oxygen you die and if you breathe oxygen you die because you are born to die! In the end oxygen kills more people than any bad habit or disease because it slowly kills everything with life. The accuracy of their basis on which physics rests is that mass is responsible for gravity by the pulling thereof. If you don't have mass you're not going to have gravity. Mass is equal to gravity and gravity applies only by measure of mass. If mass is present then it's by gravity or otherwise mass can't be absent for then gravity is absent. If a body falls it is the mass that pulls the body to fall because the body receives gravity by ratio of mass and mass is that which produces gravity in relation to the mass available.

Please take note that reading this book will to some readers seem to be **intellectually** **challenging** to any person since what I say was never yet published. I would like nothing more than to get my message across to everyone I can manage to reach. However the book in the form that this is I know comes across as challenging because the facts I present is new but not only that, it contradicts everything you ever thought was correct. For instance I say there is no big or small and yet everything you can think of is either big or small. I don't want to touch any of the other contradictions you will come to read because then you will think I am a madman that has become dangerous by insanity worsening ever minute. However when I show my reasons why I say there is no big or small you will have to agree with me.

With this outlook on science I disagree in principle with science's accepted principles on even the very basic issues and that fact is undeniably true. I also propose new principles and what you read is very new to everyone alike. Yet, the Super-Educated-Masters have preset conditions they prescribe to information and they can't break their mould. As my views are new the Super-Educated-Masters only use information stored by culture and if they can't bring the information to mind by recognising they fail to understand new science concepts. So although you will see that I am absolutely correct yet they fail to understand my work. You are going to read evidence of this in a letter that was sent to me proving just that. You will learn that I call physicists that studied for so many years just to have learned nothing in the end Super-

Educated-Masters because they think of their positions as Super-Educated-Masters but don't realise the most fundamental foundation of physics is completely wrong and yet they build fantasy castles on presumptions that does not exist. Where the Super-Educated-Masters fail to see my arguments and the obvious correctness thereof they also fail to see that I found that the ordinary persons with a scholastic physics background cope with the difficult explaining much better than does Super-Educated-Masters.

The ordinary person do not have a culture to defend and a work ethic that might be compromised whereas those in Academic office have a lot of Academic pride and years of study material they will lose. Therefore, the purpose with which I wrote this book is to get around the network by which Super-Educated-Masters strangle any form of science that does not fit their views or match their liking. If what anyone says does not stroke with what the Physicist says control physics and agree with "Mainstream Science" or echo their thinking, they just smother all intellectual publication on the grounds that it is not fitting their profile on science. I disagree most strongly but I do also supply proof thereof. Still Mainstream Science blocks the publishing of my views on science that does not compliment their views. If you believe science is more accurate than God don't read further and live out your fantasy. If you want to know the truth about **how students** and the public **are brainwashed** by **mind control** in **science** this will wake you from your slumber. Read this and wake from the culture you believe in; that which science has lulled you into and made you accept science as the absolute undeniable rock fast truth. They instated this culture concept of science only working with the truth as a religiosity. Should you believe that then stop reading or get your tranquillising anti depressants next to you with a large bowl of water. "Mainstream Science" hides behind maths. You will find some mathematical equations, if you are not familiar with it ignore it because it only shows the silliness of "Mainstream Science". If you don't read the mathematical equations because you find it tiresome you will still understand the linguistic explaining when reading the language whereby I explain it. I need help to fight their fraud. I need you to help me fight them.

Many years ago, while I was teaching, I couldn't live with my conscience and get paid to betray my scholars by lying to them about what accuracy the science I was teaching them had. I had to quite teaching. It was better for my sole to live on the brink of starvation as I have been doing and experience complete poverty than to lie to the children that look at me with those trusting eyes. To them I am completely trustworthy, even more than they believe in the knowledge their parents represents because I am the teacher and the teacher they believe. …And then I look into their eyes and tell the about science I know is totally fictitious. How can I take money and in the end meet my Father in Heaven and tell my Father in Heaven I sold not only my sole, but also the soles of children I had to teach for money. I had to live an entire lifetime with lies and betrayal not to one or two persons but to scholars year after year.

I had to teach science and I did not believe a word about what I was teaching the scholars. Then I had to force my innocent students to believe what I don't believe. How morally uplifting is it to sit with questions that need answers while giving information that you see is completely false. Ask any mathematics teacher to explain the Newton formula on gravity $F = G\dfrac{M_1 M_2}{r^2}$ and the mathematics teacher will tell you there is no chance in hell that this formula can apply as Newton said it does. According to this formula there is trillion times more gravity between my feet than there are between the earth and the sun because the smaller the radius gets, the more it increases the influence the mass has.

With all those science professors all having many doctoral degrees in science and mathematics across the world on every continent there is they never saw the error of this formula let alone all the other bullshit they were propagating…and still they believe there is no Science Conspiracy Network in place to fool the public about the correctness and the honesty in science? You wish to tell me not one could question Newton on the merits of this formula…and then they all act as if there is no conspiracy going on. Well I got to a point that I could not be part of such a conspiracy any longer and I chose rather to live in poverty than to drape my sole with hogwash. Never is there one teacher that makes an effort to explain the formula. I will never accept in all this time that not one sat back and saw this was a load of rubbish.

The context and the formula is pure rubbish and is a fairy tale. However it comes down to brainwashing, which is a culture we cannot do without within civil society. Confusing people into stupidity is civilisation because with all the perks and with all the trapping civilization brings the idea of civilization rests on brainwashing. From the start we humans never took leaders as the strongest but leaders came from the ranks of the more intellectuals. What an intellectual tells you about religion you accept because arguing with him while there is a strong case that he could be correct made the chance of disagreeing far too expensive to take. Going with the idea of what is intellectual also come this veneer of being honourable.

Where and when did it become acceptable and even come across as being intelligent when by behaving intelligent you lied and cheated and doing it with cunning then be thought of as a wise and clever person? This is when those that showed much criminal intent and unsavoury behaviour introduced a god that is dead and that is the god "money". We all bend in front of money and we all crawl before money and money is a god that is dead. On earth we fight for one god who is money and the god money finally destroys every person. When I say I am about to show how money started off then everyone starts to think of where the tale of money started in our using of it. We think in terms of what traders offered other traders and then those traders started using a system where some easier form of currency had to be introduced to make purchasing and selling commodities easier. We think of honouring a genius unknown to us today that had the inspired vision by which to introduce tokens the people could put a value on and use it as bargaining devises to ease the burden of trading and thinking this is pure nonsense. That is what the Mammonites want every one to think but the truth of the matter goes back much further than that. The Bankers take their tax or share long before the Government can but they call it bank fees. People are so brainwashed and beaten to a pulp by systematic control of the mind and their thoughts that they fall into the practise of doomed slavery without trying to fight for freedom. In the days of the Romans and the Greeks slaves were paid 10 %of what their Masters earned from their services while the Masters still had to feed, cloth and shelter them at the cost of the Masters. We all are slaves to the bankers but we earn about 10% of the cut they take.

Slavery or so I am told is illegal and banished by law and so I do believe from the bottom of my heart that slavery in any form is wrong. But the slaves did not have it so bad in the days of the Greek and Roman Empires. They were much better off than us the slaves of the current World Order. Slaves under the Roman law were fed clothed and accommodated on the Master's account. The law was that the owner of a slave had to feed him and provide accommodation for his slave. Then the slave had the right to ten percent of the income the owner generated from the services of such a slave while the slave had the chance (if he could) to buy is freedom. Slaves in the current World Empire of the Hoggenheimers and Mammonites enjoy the pleasantness of a just system where the system does away with the need to bay slaves; the slaves join the system or die. I know this because the system turned on me. Furthermore they make the slaves pay from their wages for food, logging, transport and clothes while the Hoggenheimers do not even pay them ten percent of what the Mammonites earn from their services. Under modern law, modern slaves are worst off than slaves two thousand years ago! And to top this deceased person that crossed to where the Mammonites could no longer rinse him dry had the audacity to escape the slavery without even paying his last bid for his freedom. How criminal can a man become to die in such a manner of escaping what were rightfully his dues to pay? With all the simplicity about life and the promoting of escaping death why can the atheist not bring those deceased back to do his last part and fill the already overflowing money caskets of the Hoggenheimers and Mammonites? The Hoggenheimers make the family pay for the funeral that is completely unnecessary because if left the body will disintegrate into atoms in a natural process. Just get the body in a cave and leave it alone.

It is a result of the Hoggenheimers and Mammonites wanting slaves but not to be bought as property because then it will burden their capital. It will give the slaves property value while the Hoggenheimers and Mammonites can have all the slaves they want without putting a price on the commodity. By not paying they set living standards every sucker wish to follow and then agree to give a little of the blessing for a lot of labour. If a person work such a person may afford a drink of water and if a person works hard he may afford food. If the man his wife and his children work they might afford a home but for such a privilege they have to pay the Hoggenheimers handsomely in cash. By enslaving any person's sole he may have a roof over his head. By telling the women how miserable their lives are going to be should they remain at the house and raise the kids as a housewife they (the Hoggenheimers) with that then get at least fifty percent more slaves to use and pay them with favours that belong to mankind in any case. The people can have electricity if the pay but it is not the Hoggenheimers' property to sell. People could have fuel but the fuel comes from the earth stored for billions of years and has no nametag on it and so it belongs to us all. They give them human rites and that rite they have in mind is not the natural rite to eat or drink but is to serve with being able to vote. To live they charge money because eating to love is not human rite but voting is! Democracy is a human rite. Voting is a human rite but eating and drinking is an affordable commodity that one are privileged to have because as a slave one can afford the luxury.

How senseless did propaganda become! By getting the woman into the work force they cut the salaries of the men by at least half and as a bonus they get a large group of sex slaves. The Boss at work is the Master in charge, which sets down the rules and the woman refusing him is out on the street. I know about the lawsuits and the laws against sexual harassment but those Hoggenheimers is in charge of the politicians voting for the law as much as they are in charge of the civil servants writing the laws. The laws...all laws are one huge smoke screen to favour the powerful and dominate the powerless. My human rite I have is democracy and the rite to vote. Good lord are people that easily fooled? That means I can go where another sixty million other suckers go and turn the table on politicians because I have a say but that say is one say amongst sixty million others. The Hoggenheimers lifts his pen and the politician ignore every promise he ever made to the democratic voters because the Hoggenheimers and the Mammonites will withdraw his campaign funding and leave the politician out in the cold should he the politician not play to the rules the Mammonites and Hoggenheimers lay down. The politician will have all his morals intact but he will be jobless. He either gets bought or killed. The politicians underwrite the laws that favour those in power and the laws are meant to fool the masses and protect those in charge from the masses. That is democracy at its best but it can get much, much worse with legal slavery going onto all which is much worse than the standard Chinese form of slavery as is the case in America but I do not wish to enter that debate.

While the Mammonites pay us 10 % of what we earn from what they earn from or services, we have to cloth ourselves, feed ourselves and our children while we purchase houses form the Mammonites and then find that behind successful Mammonites there are Bankers pulling strings by supplying worthless printed paper we accept as the commodity we will work all our lives to accumulate and possess. In the end we can't take with us anything we ever wanted on earth because it is worthless.

Where and when did it become acceptable and even come across as being intelligent when by behaving intelligent you lied and cheated and doing it with cunning then be thought of as a wise and clever person? This is when those that showed much criminal intent and unsavoury behaviour introduced a god that is dead and that is the god "money". We all bend in front of money and we all crawl before money and money is a god that is dead. On earth we fight for one god who is money and the god money finally destroys every person. When I say I am about to show how money started off then everyone starts to think of where the tale of money started in our using of it. We think in terms of what traders offered other traders and then those traders started using a system where some easier form of currency had to be introduced to make purchasing and selling commodities easier. We think of honouring a genius unknown to us today that had the inspired vision by which to introduce tokens the people could put a value on and use it as bargaining devises to ease the burden of trading and thinking this is pure nonsense. That is what the Mammonites want every one to think but the truth of the matter goes back much further than that.

This is the Mammonite. He is the one that has money and therefore he is the one that is clever. He is the one that has money so to him a politician is a pre-paid investment there to serve him and bullshit all others. He is the one that has money and he pushes everyone else off the table not with a swipe but with a glance. He buys democracy that people vote for. He feeds us by "*providing*" people wit jobs. It is not said that he uses slaves he did not purchase but he bought a treasure of slave by having a bank with a vault. He tells the politicians and the clergy and the scientists to do what he says. The Mammonites finance wars and we see this from before history was written.

They are the ones who can see in the dark and they are the ones who can see through your bedroom window and they can visit your wife when you sleep and they can farther your child when you are not looking because they have the power of God on their side by practising dark Magic. We know we are cheated hands down by government but it can't be because of our own impotent stupidity. We know we are cheated politically but they must have an evil Satan Devil on their side to play with us so ruthlessly.

Everyone wants change because what we have at present serves the Mammonites by feeding the populous to their riches. Every politician shouts about change that everyone wants because we are at the mercy of the Mammonites. George Bush being the President in office promises change during an election campaign while he serves the rich. He sits in office knowing the public is unsatisfied with the way things are but instead of changing while being in office he promises to bring change. No one wants what is coming their way and still they want to be part of the process that rules them, so they opt to go democratic.

We think their power they derive by which they control all aspects of our lives can't be because we are brainwashed into thinking democracy gives us power. No, if that is the case then we must take the blame for being that stupid and we are clever since everyone in power tells us we can fight for what we want because we are committed to democracy. We think that it is by democracy that we govern the country by putting government in power and we tell the politicians how to govern because we can vote, and we are the ones living in the 21 first century thinking we came to be the most intellectuals of all times? In Britain there are 44 million voters and of the 44 million each one can vote once.

To cast my vote being a believer in democracy I have to stand in line sometimes for hours because I am one of that many fools just to make a cross next to some political party and that turns me into Superman while in fact that makes me a bigger fool than the Neanderthal Ape-man was when they ruled the earth during the time before he could think because we think we can think! If you can jump off a cliff and I tell you every once in 44 million times someone lives the fall to tell his or her tale, will you jump? Still you do think being one of 44 million, that honour to vote gives you the power to rule the country by democracy? Can we still find people that stupid, yes there are and the western civilisation is filled to the brim with such mindless masses, begging for the privilege to be able to vote. No wonder those in power can keep so many on as slaves and never pay a dime to have slaves to fill the sole purpose of making those in power also rich at the same time. …And who is in power, not the politicians or the clergy or the King but the low life Mammonites that has banks and with that the power to purchase the services of everyone than money enslaves including me.

If I cast my vote I can change the world. What a bunch of dumb bastards and mindless idiots would fall for that. To think if I was a Britain I have one chance in 44 million and still they want me to be an idiot to think that the one vote I have counts. No, I can't be that stupid in believing I am played like a drunken chicken! No there has to be an evil-eye bunch of Devil worshipers who call on the power of Satan by formulating dark magic rituals and then they get the power to influence my life so that I become my worst nightmare. I become so stupid that I believe it is my right as it is my duty to send my children to be killed in a war they wanted and planned and that they put in place so that they can profit from and to protect their billions. Fortunately for me I now am in South Africa ad now I am pert of another 45 million idiots and I have to stand in the sun for hours on end burning my brains out and wait for a chance to cast my vote that has no chance to be counted because there is no way in hell they can count 45 million votes in three or four days. The best is that every sucker is brainwashed to believe in the "system" of justice!

Hey liquid brain, you are not one in a million making you special, you are one of 44 million making you anonymous. No matter whatever the party is that you vote for, the Bankers got to the politicians first. To thank Tony Blair for giving the oil fields of Iraq to the British oil barons, they awarded Tony Blair with $30 million this far just to go around and make speeches. He gave the bankers the oil and while he is alive he is forever a rich individual. It paid to allow the Liberal Democrats to be against the war to present a democratic resistance and pay them handsomely for being negative about the war, while the Liberals and the Conservatives believed whole heartedly the photos of the weapons of mass destruction was real and that piped out to be demolished war debris laying in the desert. So both the main parties voted for the war that would enrich the bankers and industrialists much more than they are already rich and the politicians on all fronts were paid generously to believe pipes in old mines could be atomic missile launchers. All they had to do was to brainwash the public about the urgency of going to war to save the country!

No matter for what party you vote, you will keep the money-mighty Mammonites in power and in turn they will keep the Politicians in party as long as the politicians serve their cause well. The public will still have the privilege to send their children to death to become national war heroes and die for national pride while when they get back there is not even money to supply the war cripples with artificial limbs or to give the crippled a living pension. The penniless have the fortune to send their offspring to die and to murder so that the rich and powerful can become richer. When will Tony Blair's boy go to Iraq or Afghanistan to serve his country? Is Gordon Brown's two boys going to enlist to fight for their "Queen and country" No it is the " grass roots" scum that only has a purpose of keeping the rich safe that would have the honour to be killed for Queen and Country or to kill as many Muslims in Iraq as possible for "Queen and country".

Turning our heads to America things are much worse because in America I think you as an American will compete with something like 160 million to one to get any change in the two houses Governing or to get the President to go look for a new job. Think of what the chance is to fight a battle against things you don't care for and playing a 160 million to one role in the outcome. You lot democrats are simpleminded. Where did this knifing all begin that makes us slaves of those that holds money and power?

I wish to begin this by beginning to explain the biggest swindle man created. It is called money and in truth there was never anything such as money. Money does not exist. To have money one must steal from the earth and then swap this for merchandise others lawfully produced. Only by taking from the earth that which either belongs to no one or to everyone alike and illegally claiming ownership of what the earth produces can you give money any credence and value. The idea that money can form a basis as a visible valued factor is a hoax. Money is only taking what the earth holds and then trading that for what others have. However to put say coal into the trading system can only bring currency once and then the theft is converted to money.

There can be no repeat of creating money because the theft happened once and the purchase it underwrote happened whereby the gaining of money is suspended into vapour mist. The coal served a purpose by coming into a system where it never had a place and it loses the place immediately by finding legal ownership afterwards. Say it was traded for a cow the cow carries the currency forward because the meaning of the coal is lost as a bargaining chip that others want or need. Then as the trader falsifies prices by raising the value of the trading commodity of money could money find further value and the truth is the process is lowering the value of the goods purchased or sold on. There is no honest manner that can make money more expensive except making goods less valued. It is more robbery by enlarging the role of the thief. But that is to the detriment of lawful commodities owned by honest people. The liars and the no-good-cheats are those praying to the money god Mammon and are named Mammonites after the god that they live for and the god who they uphold. Some call him Satan, others call him the anti-Christ or the Devil and in the Christian religion the Messiah was crucified because he fought Mammon by throwing out the offers that was for sale, turned over the tables of the money launderers and threw the money into the street as He whipped the merchants. Go on read the Bible and you will see He was crucified because He threw out money from the Synagogue and after that the Jews took Him to court. I guess every religion knows this Mammon by another name the assimilation is representing evil. Yet I try to live without money because I am forced to and there is none tougher task.

I never discuss me religion with anyone outside my family but after some sole searching I came to the conclusion that the following is not religion but is the science of corruption and I am not going into religion as such other than showing how far money corrupt society. Again I press the point that I never try to per sway or to convert anyone to my religious devotion because I believe whatever you believe is going to save you and if I explain anything further than that statement I am going to go into religion. I would just say this: converting anybody to any faith is a waste of time because nothing is more exclusively personal than religion. Still…if you are not part of Christianity read on and see how clergy corrupt to further the ends of Mammon and money. This is a factual case and has nothing to do with belief or conviction.

If you are part of the Christian faith what I am about to tell is part of the most basic idea about how Christianity is conveyed by all denominations and faiths there are in Christianity. Every body preaching Christianity will tell you that Christ died on the cross for you sins and to save your sole from everlasting condemnation. If you are religious believe Christ died on the cross for your sins and to show your sincere appreciation you better give the church money. Giving money will show your dedication and gratefulness for Christ dying on the cross to save your sole. Yea sure, and don't forget to bring money to the service just to show God how thankful you are for His goodwill in letting His Son die on the cross. The cross is used as a symbol to milk your feeling guilty and it is there to remind you to feel guilty that someone else bought your eternal rescue by hanging on the cross. There is no better way to get the money flowing from everyone's purse than to see that icon and become filled with unbearable sorrow, endless guilt, borderless appreciation and overwhelming gratefulness. So this is what you will find going to any Church that teaches the teachings of Christianity. Well this is the one part they want you to believe because with that the money is rolling in! …And now I am going to give the true story and you can go and read the Scripture without blindness and become wise. Remember firstly I couldn't care a buck-f#&*k about what you believe or don't believe and what's going to happen to you after death. You have your worries and I have mine and I am not there to convince you about mine because I carry only my responsibility.

If you believe that Christ died on the cross then this is why and if you don't believe that it happened read on and learn from what you think of as a fairy tale. I am not trying to convince any one ever. Christ did die on the cross and yes he was charged but Christ was charged with another crime other than you having sinned and He had to pay the price because you sinned. Christ overturned the tables of the merchants in the synagogue and he whipped the Mennonites into the street. He let lose the birds people had to buy for offerings and he chased the animals away. He completely demolished the merchandise and the profiteering moneymaking dealers and screwed the entire market that was part of religion. He was charged because the Rabbis would not stand for such ruddiness. Teaching rebellious ideas was one thing but to destroy the commerce with malice was one road too far to travel and that had them taking Christ on a criminal charge to face the Romans. He was charged for breaking up the money racketeering by throwing out all the commercial entities and counteracts the wheeling and dealings going on every day.

For that he had to die on the cross. With that act he freed all future Christians from bringing money offers to god to find absolution. He disconnected religion from money and from that onwards every believer had a direct line in praying to God and did not have to bring offerings so that he or she could worship. He dissolved the connection there previously was between money and religion. From that moment onwards Mammon stood in the role as the Anti-Christ because Christ threw all money out of the synagogue. Every person excluding no one that brings money into Christianity is part of the Anti-Christ because Christ demolished the connection there was between money offering and connecting to God.

Christ died because he got rid of money so that money no longer presents any function in Christianity. He died to free me from my sins but that was to get money away from religion and banish money from the Church. He was charged and He died because he threw out the money and the merchandise. Yes, now I can find absolution and have a connection to God and now I have a free pass to God. Will any clergy agree with me, no never and if Christ came again tomorrow to banish money from Christianity all Christians carrying the cloth will crucify him on the spot because no one dares to touch the money that religion generates. Take away all the money and the properties from say the Catholic Church and see how they nail that person to a cross with or without a hearing or a fair or unfair trial! Anyone in argument with this I wish to present the case of the rich young man that came to Christ and asked how the young man could get to heaven. Christ told the rich young man to sell his properties and rid him from all his fortune and finance. Christ told the rich young man to give everything he had as wealth to the poor and come back as a penniless beggar just as Christ was. Christ was the poorest of all that walked the earth because Christ had no place he could ever call his. Christ said the jackals have holes and the birds have nests but the Son of man had no place to call home. Christ did not say to the young man to get baptised or to start believing in this or in that manner or follow this or that denomination or become part of any specific faith! Christ did not say to the young man to liquidise his wealth and bring to bring the money to the apostles so that they could have more comforts and a bigger paycheque at the end of the month. Christ did not received money so that Christ then could baptise the young man and then bless him eternally for bringing the church or the congregation such a lot of money. He told the young man to free his sole from the money that burdened him and give the money to the needy. To get to heaven you can't have money because money is the dead God or the antichrist or Satan or whatever name you connect with what turns out to be unwanted. This is also said in another text in the scripture and it reads using as many words. It says rich man you cannot enter heaven because of money. When the rich man showed unwillingness to part from his money Christ had tears running down his cheeks in sorrow because Christ knew that the young man was lost to Mammon. Why did He cry if the young man was not lost to money?

So many persons challenge me on this argument but it again is not me trying to make up things. I say what is said in scripture and like it or not there is a choice between religion and money. It is said in so many words that you can't serve two Gods, Mammon or the living God for you will forsake the one and worship the other. Ever sat in a church and listened to sermon in a two hour service where the pastor or preacher is carrying on relentless on that topic. Ever sat and had the meaning behind this serving two Gods and distancing you *faith* from money and to exclude any thoughts about money in terms of your faith and where this is explained in the most detail? No, you will sit for hours and listen that it was your fault Christ died on the cross and that you should shamelessly take the guilt because of what you did to Christ by sinning…and by the way leave a few Bob just to show that you are very sorry until next time you have to buy guilt relief again. I am telling you in the mode and in the manner that Christians run their faith and their scruples at present that if the young man came to any faith the man bought his everlasting salvation the second he handed over the cash to any clergy in any congregation of any Christian faith.

Christ died not for your or my sins but because he threw money into the street and whipped merchants by chasing the Mammonites down the street. Christ banished money from religion and for that He died on the cross. Have you ever heard of this detail? Never because that part is unfit to have as a part of religion. Everyone serving in whatever capacity or form of Christianity is serving Mammon first by collecting money and then sells a service in prayer of spiritual comfort but that is in exchange for money.

Then you are going to tell me what I presented is pure religion. No, its not, its pure corruption going as far into the Christian faith and as deep as there is a Christian faith! All clergy from any denomination will perjure the highest form of Christianity to sell Christ out to the antichrist so that they can serve Mammon well. The conspiracy to hide the truth about why Christ did die is hidden so far down by all serving the Christian faith because they sell your sole for money every time they get you guilty on what you are and that you have to pay (more in money than in anything else) because you are what you are when you are who you are! They conspire to defraud all Christians in the name of money and we all play along.

I'll give you a precise example whereby you can obtain an insight into the mindset of the Mammonite and the way the Mammonite thinks and functions. I am a South African and being white being poor means having another definition about poverty than some of my black countrymen. I am dirt poor, as poor as a white person could ever get anywhere on earth but I have always got a meal ready to eat at mealtimes and a roof over my head because I have an educated wife that provides for our basic needs. I must say this at this point just to clarify my lifelong marriage. The only clever thing I ever did was to marry a clever wife and for as long as I live I never had to be clever afterwards because she takes care of all the cleverness I need. With her providing the cleverness, as she is the clever one I can philosophy as I do in this book but that is philosophy and not being clever. My philosophy up to now never brought in one cent because; well no one pays for information any more. It is freely obtainable from the Internet. With me also on top of everything being poor in health and one heartbeat away from the grave we have no room for luxury but in our daily needs we are provided for. In the black community in South Africa there are those living in squatter camps that has little or nothing to eat.

These are mainly blacks from Zambia, Zimbabwe and Mozambique because the South African blacks living in desperations are entitled to food relieve and a tiny cash supplement giving the needy some income and their desperation also carries a different definition to what the blacks from Zambia, Zimbabwe and Mozambique experience. In order to bring relief and help to the desperate coming from outside our borders some Hyper supermarkets supply large containers in which people can place food they purchase extra and then donate that extra to the needy. This kind hearted Supermarket management then promise to take it upon them to deliver the donated food to the squalor camps and bring relief to the hungry living in desperation. So why are they Mammonites, because after all they form a channel through which food are distributed to those with food shortages. You see those I mention as the most generous Supermarket management have these enormous bins in which the bleeding hearts can put the food but the bleeding hearts purchase the food they donate at full shop prices.

This takes us back to the bleeding heart baying off guilt by paying the philanthropist to collect on behalf of the Hoggenheimers dishing out to the Mammonites paying the Mammonists for some slave driving. The actions are deliberate but the true intentions are deliberately unintentional. We are bullshitting our conscience for gaining our mistrust. It is the lye of culture and all participate but some participate to a degree that does not please others. The degree might be to some extend not serious enough to bring commitment and the persons would stand on the side line and criticize without direct involvement because of fear of own guilt uncovering or even of a want to participate while others would come in and rescue but not to save but out of spite because of personal yearning for participation that the person knows would not be permissible. If we give is it to feed our narcissi or is it to serve our master Mammon?

Donating in any form becomes the Sudan affair where the bleeding heart buys guilt relief and become god to those lesser while the philanthropist pushes guilt as hard he can and be god to collect money on behalf of the Hoggenheimers that then can be god with such wealth distributing it to the Mammonites who can be god by buying from themselves as much as selling to themselves with unscrupulous profits making him god and allowing the Mammonists to be a slave driver and being god to the slaves. It is this sickness of society no one cares to see because every one gets what they want, even the luckless get what they want with the minor condition that when the luckless suffer most that is when it becomes the region where most profits are for every one in the chain of gods. So the luckless must be in crises starving as they are dying to gain most profit for every one. The profit has little to do with money but with being god. Any attempt to stop the situation will never be tolerated by any party and therefore my remark that every one will press for my castration because of my suggestion to rectify and bring a solution.

There is no line or special till that you can run the donated food through where the till will give discount of any sorts towards helping the public to donate an even larger amount of food. If you buy even in the event of buying for the needy they see to it that you pay full price and in cash! Those being the most generous Supermarket management ask full price for whatever the bleeding hearts purchase and then they make a large profit on the sympathy of the bleeding hearts as well as the desperation of the hungry and needy. To them being the most generous Supermarket management the channel to provide for the needy is only an outlet to encourage the bleeding hearts to buy more and therefore push the most generous Supermarket management's turnover profits higher. The Mammonite takes advantage of every situation to enrich the bank coffers of the miserable cheats and line their pockets with larger sums of money. They urge others to purchase and donate towards the needy so that they can sell more and make more money from all other person's needs and kindness. That describes the Mammonite. How did we allow these bloody bloodthirsty bloodsuckers get their paws into our society?

This takes us back to the bleeding heart baying off guilt by paying the philanthropist to collect on behalf of the Hoggenheimers dishing out to the Mammonites paying the Mammonists for some slave driving. The actions are deliberate but the true intentions are deliberately unintentional. We are bullshitting our conscience for gaining our mistrust. It is the lye of culture and all participate but some participate to a degree that does not please others. The degree might be to some extend not serious enough to bring commitment and the persons would stand on the side line and criticize without direct involvement because of fear of own guilt uncovering or even of a want to participate while others would come in and rescue but not to save but out of spite because of personal yearning for participation that the person knows would not be permissible.

When economy as a trading device started many years ago there were those that was willing to work and those not so willing to work and today these two groups are still with us dividing our society into two groups. The not so willing to work was thieves and low life and those willing to work were those honourable men around which culture and the future pivoted. The two groups are still amongst us with one being the working class and the others being the Mammonites, those holding money as a religion and praying on the rest to steal and to gain from what others work hard for. The grouping has no connection to religion because there are Jewish Mammonites as there are Christian Mammonites as there is Hindi Mammonites as there are Islam Mammonites and those religions I left out also carry the same ratio of robbers, as there are honest workers amongst them. Mammonites create a religion called serving the god called money and this is the true religion they serve. It is not a reference to a group of people or a renaming of a certain religious conversion but to show whom amongst us are the parasites.

When going on a discovery to find the root of all that is evil by representing dishonesty in all of man or women I have to begin where civilization began and that was (I supposes) just after language became a tool of progress. Everyone was a hunter and a gatherer and killed to live. Then the more intellectual saw that by not killing but herding animals they could prosper and find a more easy life. Farming did mean being busy most of the day and guarding the heard at night but better security for the goodwill of their children made it a worthwhile option. Some hunters saw an advantage in not killing everything but started caring for some animals they caught and looked after. They found a way to protect the animals by guiding them to better pasture where in the minds of the animals they were better off when being protected from scavenging wolves. Others saw fit to put specific seeds in the earth and harvest the crop at the end of the summer. Some saw this too be more cumbersome than hunting and felt they could become parasites by dishonesty, which is better known today, as a merchant's cunning trade ability. It still is robbery and it is dishonesty. The merchants were too god damn lazy to work and swindled the honest person by cheating.

This also meant by farming with animals or crop they were attached to the needs of their animals and could not roam as freely as those that chose to hunt but hunger was less prevailing when more effort was applied to look after the animals that they herded. There was not much free time on hand but there was job satisfaction in the price they paid. The rise in numbers of the animals brought more than their needs asked for so a surplus arrived which stood them and their children well as they could prosper.

However those not willing to work had a lot of free time to gather what the earth provided and could harvest what the Mammonites never planted. They could steel from nature and trade it off as property owned. Those not working could gather from the earth things that were not theirs to gather but they took it from the earth anyhow. They got hold of flints and metal and things the farmers needed and the farmers had animals the Mammonites needed to survive with in the harsh icy winters when hunting was hard. Someone had to have a plough, someone had to have an axe, someone had to have a knife, and someone had to have spear points. So, people had to have something manufactured. When everybody had animals, the one had a cow the other sheep or chickens and then maybe said the cow was worth a hundred chickens. So, they started trading. A cow was traded for ten sheep or a hundred chickens. Everybody was happy.

It was a trade off. But no one realised anything. There was no growth. It was just trade and everything was resolved by swapping. Then someone at the very beginning discovered flints and thousands of decades later this became copper and then iron. Now there was a farmer who had ten sheep and he wanted an axe, which at the time was a wooden handle with a large stony flint tied to it by goat guts. So the farmer needed a flint to chop wood so that his wife could cook food. The brilliant clever or cold hearted criminal depending on who's point of view we find a description got a flint for free and brought this to the farmer that worked his arse off for one year to produce what he harvested albeit in animal form or in plant form. The farmer being honest thought that the criminal merchant also toiled for one year as the farmer did to produce a crop as a product and with that in mind had an idea that what the criminal offered had much value. It took the farmer one year to produce one cow and he traded this for one flint because in his honesty the farmer did not know any better.

This enriched that lazy filth forming part of our society as the merchants. The Mammonites never had to work but only pray on the needs of the honest and then rob them blind because the working class were the honest and sincere group. This carried on while the dishonest Mammonites prospered to begin becoming the rich Barons and later Kings and forced to honest people to submit to the Mammonites' gangster enterprising. The Rich criminals became the powerful because the dishonest purchasing brought the imbalance that still prevails. Any Hypermarket buys from the farmer what it took the farmer one season to produce and then this the Mammonite sold to others for ten times what he pays for it while it takes him hours to sell from the point of purchasing whereas it took the farmer on year to raise the crop he sells for one tenth of what the criminal sells for. How do I know all this; I was the farmer and I sold to the merchants in this manner. This practise of cheating and plundering became commercial progress and all honest persons suffered equal at the dishonesty of the criminals that set up shop with enslaving the honest as their main intent.

Let me explain the difference between the mindset of the honest persons and the thinking of the Mammonite furthering his criminal intent. Where it began in the beginning when money became a tool of the oppressor to construct a basis for the rule of the false and dead god the farmer had to plant crop to find sufficient supply to cover his needs and the needs of his family. Before farming became a structural enterprise was to go into the wilderness and locate fruit and then harvest the fruit but this practise took up too much time to make the effort viable. Stealing from nature was no option that leads to survival and therefore harvesting the seed and preparing field in which to plant sufficient quantities was the only way to go. If the farmer wished to survive from the trade he chose he had to put in time and labour to harvest enough to allow his family to survive. But enough was also to harvest more than was needed for personal use because one can't live on bread alone. So he had to harvest the crop he chose as his "bread" to find some "butter" to add as spice. He had to harvest beyond personal need to trade with other farmers whatever they had as farming produce in order exchange what he had for what they produced to give his family and the family of all other farmers a balanced diet and in this manner everyone found a balanced diet by eating more than "bread" alone. Self-supply extended beyond personal comfort and went to providing for the need of your neighbour as well and in that trade became a manner of being civil. That good heartedness made way for the plunder of the Mammonite as it does in current times.

There were other basic needs the farmer had to employ as tools. The farmer used an ox to plough but while using the labour of the ox the farmer cared for the ox by providing fodder and water in order to tend to the ox because the good health of the ox was detrimental in supplying his family with essentials. The farmer could not train a new ox to plough every year because it takes years to train an ox to do what is intensive labour on the part of the animal. Because of this he had to put a retired ox out to pasture to use the old ox to teach the new ox how ploughing was done. It was a long-term investment and generations of oxen were needed to sustain the present and all had to come from the intelligent planning of the farmer.

Sure the Mammonite also required intelligence but that was being shrewd and cunning and always planning to steal rob and swindle. If the fisherman wanted to fish as the farmer did on a daily basis the fisherman had to cast a line and pull in a fish. This made the fisherman harvest what was in excess in the pond. But then the Mammonite-syndrome kicked in and the fisherman thought it would be better to go to the middle of the pond or lake and harvest more fish over there. Fishing from the side meant the fisherman had to supply some food at a certain spot as to lure the fish to that spot every day and those fish that didn't get caught got fed along the way. If there was more fish he could catch at any given point then it would be in the centre of the lake and going to the middle of the water brought more fish but the harvesting had to use a method that required money. Why is that you may ask?

To get wood to build a boat the fisherman had to get a woodchopper to get him a tree down. If he chopped the wood by himself he could not have time to catch fish and his family will then starve from hunger. So he employed the services of a woodchopper. The woodchopper needed an axe to chop the tree to get the fisherman his wood with which the fisherman was going to build his raft with which he was going to fish in the middle of the pond. Therefore fish had to be traded to get a big enough flint to do the chopping of the tree and this brought about collaboration between trades because the fisherman had to catch enough fish to give the woodchopper to feed his family while he was chopping the tree and also the fisherman had to fish in sufficient quantities to give the Mammonite what the criminal wanted for what was not his to begin with.

The next question you may ask is what is the difference between the sleazy Mammonite taking from the earth flint that is not his to take and the woodchopper taking wood from the forest which by all account is not his to take? The money par comes about when he can take as much as he needs for personal use but not to swap it for other commodities because that is creating money and plundering nature. He will inevitably plunder nature because when he takes for trade, greed kicks in and lust for wealth makes that his consciences suffers and the penalty that must be paid for such lust and greed is the earth being demolished. The difference there is between what the woodchopper does and what the Mammonite does is no difference as they both plunder nature and take what is not theirs to take. This practise will engage the intervention of the criminal Mammonite since the woodchopper will find some Mammonite that will buy the wood for next to nothing as long as the woodchopper delivers it to his shop's door where the Mammonite then will sell the wood at an extravagant price because everyone perceives this criminal bent on enslaving others as being "clever" and "bright". The Mammonite rapes honesty and then feels confronted if anyone calls him by the trade of his character in being a criminal.

The trees are not theirs to take and when they "have" to take it then don't turn it into money by giving the tree a purchasing power. The "taking of what is not theirs" and "turning that into purchasing commodities" is what creates the money aspect and the money kills the tree by having the trees become merchandise. Britain once was one big forest with trees so abundant there had to be woodlands no person could enter. It has been desecrated by human enterprise to a point where it now is largely savannah. Everyone plundered forests at will and nature could not replenish in the numbers humans stole and now wood is shipped from South America and Africa where these two continents are turned into desert, and yet the Mammonites are with us still, ruling our politicians that we supposedly chose by democracy to rule us but instead they rune us. If you think how Darius the Great ruler of Asia in antiquity and his sun Xerxes I built on two separate occasions during the two Persian invasion of Greece a sea crossing formed by wooden rafts the wood they plundered changes all of Asia from a forest to a desert. The wood used in times of antiquity seems to have changed mighty forests into what we now have and that is never ending deserts.

As time progressed many diversions of farming became practise. One such an aspect may serve as a very good example to show the influence commercial intent had on civil development as time went on. The farmers planted vine to grow grapes. The grapes were exchanged for whatever the vine farmers needed to live until the next crop was ready for harvest. The vine farmers liquefied some of the grapes to produce wines and the wine then was exchanged throughout the year for commodities that would sustain their needs and livelihood. Then came the Mammonites and they saw what they call business opportunity as they became wine merchants. They bought wine from farmers and sold the wine on at say only ten but most probably hundred times the price they paid the farmer. The farmer had produce that was bought. The wine merchant exchanged the produce for money by elevating the worth of the commodity and by this corrupt enterprise created money that in principle is and was absent when the farmer sold to the merchant. The Merchant added nothing in terms of labour or any inset in multiplying existing quantities but elevated the prices to create money, a means and a commodity that does not really exist.

If the fisherman or the woodchopper had to produce a tree and then to produce from this tree a boat by which they could go fishing the fisherman and the woodchopper would starve of hunger because it takes from thirty to three hundred years to grow a tree that can be used to build a raft that can be used to increase the income of the fisherman. So they do what any corrupt-minded, un-scrupled Mammonite do and create money by stealing from the forest wood that belonged to no one or everyone alike. They took from the forest, devastated what was in the forest for all in time to come, traded nature for money that does not exist and messed up everything just to produce money once and with money everything else disappeared into nothing. This way of civilising civilisation brought progress and progress decimated civilisation by creating a need for merchandise that contributed to more plunder because creating money created a required sustaining of plunder, of which mining is a result practised today as it never was before and now in this text serve as another very good example to show what they take what is not theirs.

Progress brought about the mining of metals. How can this be dishonest? Then there was a farmer who wanted a saw and there was a farmer who wanted a plough. And there was someone who had iron. At first it was flint but someone got hold of copper that was then replaced by iron to become the marker for the Iron Age. This applied in the flint or Stone Age, went on to produce the bronze in the Bronze Age and gave a name to what then became the Iron Age. I fill this in to show as humans progressed all humans carried the burden of what the Mammonite-mentality taxed honesty through all time that formed civilisation. But the iron was not his. He sold what he took from within the earth by recovering the currency he used as trading merchandise by recovering iron from the earth. That was how money started. The currency started with things people took from the earth that belonged to the earth and that person's claimed what was not theirs to claim. The Mammonite decided that the flint from which one could make an axe is worth a cow. The farmer that had no axe and wanted the axe was then one cow poorer and this dishonesty was securing the value of money. He swapped the axe for a cow and the merchant gave a sword to another farmer in exchange for five hundred chickens. The cow now was worth five hundred chickens if you wanted a sword delivered by the Mammonite but it was worth five cows if you wanted an axe delivered by the Mammonite because a cow use to be worth a hundred chickens until the Mammonite got the help of the King and taxed all purchasing of essential commodities and with the help of the King and his clerks the prices moved up faster than the farmers produce or could trade.

In the hands of the farmer one cow was ten sheep that was hundred chickens but when the merchant entered the trade market five cows was hundred sheep and that in turn was five thousand chickens when bought at the merchant and which was very legal and not only was it legal but it was no longer legal to buy from the chicken farmer or the sheep farmer or the cow farmer but it had to be purchased from the merchant just because the merchant was in cahoots with the King and the highway bandits alike. So the highwaymen became tax collectors while the Merchants became advisers for the Church because they too came in on the act. The Church and the priesthood were trading in soles and in sins. There were soles to buy and sell and clemency the people had to purchase so that the Devil would lose the fight for their eternal soles.

Farmers became venerable because of criminal gangs becoming the King's tax collectors and while also collecting taxes they plundered the rest for personal profit. In order to protect their interests the farmers needed swords. So the farmer gave in to the weapon merchant's hugely inflated prices to obtain a sword. This forced the other farmers to give the merchants a top price in sheep or goats because if the farmer didn't have a sword and an axe, jobless marauders would kill the all the farmer without hesitation because the merchants paid the bandits to raid the farmers just because it was good for business. Today this is called the insurance market. Now in view of overwhelming danger the farmer decided a sword is worth a much as twenty cows because the farmer needed protection and that service was part of the global trading practise. But everybody knows that a cow is worth ten sheep and a cow is worth a hundred chickens. By starting to promote violence against those unarmed farmers not willing to pay exuberant prices the hypocrisy of the traders became a battle to arm the bandits for free to get the farmers to pay for weapons and that the farmers had to do since the farmers were desperate for means of self protection. So a sword became five hundred chickens and a sword became 100 sheep and it could fetch ten cows at your local merchant. If one farmer is willing to pay up then the rest must follow the inflating trend of prices and shut up because plunder and theft are the way the economist work. They employ the Kings guard as much as they empty the jails ands get the jobless scum with small minds to commit murder and supply them with booty. They take what is not theirs to take; albeit stolen goods, looted goods, King's ransom and miners plunder it makes no difference because the lot represented money. This had to be money because if it were legitimate goods it would have had a trading value where one person that has something he harvested gives someone else something in return for something the other person had to swap. We see this with the trading of oil, food and electricity and the mining of gold even to this day.

Then there the notably honourable economist had to go dig in the mountain and retrieve more iron core, which they melted and produced metal and started to provide everybody with iron and start a civil war. This was labour intensive and we know the lot that did not go farming was too lazy to farm and they then will also be too lazy to mine! Now came the point where the economist had to perform in much brilliant wisdom. Right now we have the economist ruling the world by paying the politicians to help with the plunder of the commoners but back then it was only the King they had to approach with lots of money.

But lazy as they are while they think it is intellectual the economist-Mammonite got others to do the digging and do the labour so that this Mammonite can feed on others labouring. This was a bloody good enterprise. Those the King did not kill in some war that made the King feel as if he was equal to God could die as young as if in the King's army by mining for the Mammonites that was serving their god they named money and was well fed on sheep-shit and cow dung because the miners had to repay the Mammonites for such kindness and sacrifice as performing job creation. The Mammonite gave the gold-digging labourer sheep-shit to eat and expected gold billion in return. The sheep-shit the Mammonite got from the farmer for free so that the labourer came living and working very cheap. To pay for the minor's labour the economist made the miner purchase the sheep-shit at a company store and made him pay through the roof for what the farmer threw away. This is admirably clever and cunning financing practise.

The cow farmer had his sword, the sheep farmer had his axe, the chicken farmer had his metal piece that he wanted and the metal piece did not run away. The economist with his iron had a problem. Trade went slow because the merchandise did not move and the need for purchasing power did not grow as rapid as the merchants grew in numbers. So as all economists do, he went terminal. The people who did not have cows or sheep or chickens, the economist told them that they could have the metal piece they wanted if they brought him 2 cows, 20 sheep or 200 chickens. Then those that had nothing could have the sword. The guy had no animals. So he went and robbed the daylight out of the farmers. He brought this guy his 20 sheep or 2 cows and he paid for the sword. So money was created in terms of robbing those that have and sell to those that want to buy but do not want to pay. Now, suddenly the sword was worth 2 cows, 20 sheep or 200 chickens as long as the one that had nothing had animals with which to purchase arms. That was the first time inflation or growth was created. He who had nothing stole from those who had something to give to those with criminal minds also known as criminals or economists and the economy started growth. The economist had those who had nothing on his side, supplying them with weapons to kill those who owned something to bring to him who had the metal factory and when this happened we had the first bankers. In all of this the King was paid to guarantee protection in order to maintain free trade amongst those that had nothing but always had something to sell. Today we call them shop owners. The bankers have animals deposited at their doors for whatever merchandise they had available and whatever they had was illicitly obtained from those that had nothing but always had something to sell. Now the banker or economist had another group of people who did not want to rob the farmer, kill the chicken farmer but had to eat. So the economist gave those people labour. The economist had to get something in return. He put them in the mine to work. And then at the end of the week he gave them a chicken stuffed with sheep-shit as a bargain just to keep them alive to retrieve iron core for another week.

He gave them just enough so that their children will starve and go hungry but the workers will stay alive. Seeing their hungry children starve was motivation to work harder to enrich the Mammonite even more and that spurred the Mammonite of to let the children become ever more desperate for food and that spurred the workers on to become more loyal and work harder for less benefits. In this money surplus grew. So, the biggest criminals on earth are the economists. They kept slaves to feed their economy and had the criminals who kept the economy alive and they had the worker which they gave work by job creating to retrieve the iron core, which was not the property of the economist or Mammonite to begin with. The more iron miners retrieved the more money there is because it becomes money and the faster the criminals steal from the working people and then the working people had to purchase again to replace what they wish to protect but what they will lose in any case in the next robbery the criminals do on behalf of the economists. To be able to keep the economy to flow the economist had to keep on retrieving from the earth with mining to create money. This they did by stealing from the earth, that which wasn't theirs.

And they swapped it for farmers that farmed the cattle and there was a normal flow of increase in cattle. Only when you retrieve that which is not yours, you can have money. Otherwise you have equilibrium. The one will trade his cow for ten sheep, tired of eating cow meat. The other one will trade a pig for hundred chickens, because he is tired of eating pig. But there is a normal swap where nobody loses much, but nobody gains much either. The one that can go into the bush and shoot the pig and came back and traded it for a chicken, the farmer then did not eat his own chicken or boar. The price remains the same – therefore equilibrium. But the hunter living in the woods needed iron. He needed a bow and arrow to shoot the pig. So, he started trading for little. The miner got for nothing because what he used as currency he was stealing from the earth. And he was using people to mine for him…because he was too god damn lazy to work. Then he decided on money. Money is the root of all that is evil.

Now there was a prince. The kind ruled over all these peasants. But the king saw another king having a kingdom of his own. And he had thoughts to fight the kind and gain his property. Then he decided to buy swords. But to purchase these swords he had to do what the merchant wanted. The prince thought it wise to listen to the Brainy that had insight into forces that pulled and pushed weather conditions, could calculate and command lightning and could order floods just by studying the stars. It would be much better for the Prince to have this lot on his side than to care for the beneficial conditions of the hard working farmers that was already over burdened with taxes and contributions to the Crown and every war effort the Prince could dream up. If the farmers started to get grumpy the Prince sent his army in and pacifies them into silence and obscurity with sword wheeling and knives cutting flesh to the bone.

In this it is not hard to see who the Prince would favour to carry the banner on his side. These Soothsayers and Alchemists could order a fire by instigating a lightning flash and set the thrown on fire while the common farmers did what they were told and did it without quarrel. Could these wise wizards do what they said they can…no but it was better not to take a chance. So the Prince took from the workers to feed the Scientists that could see forces pulling and forces pushing and Alchemists that could turn lead into gold and paid the farmers money while taking merchandise in the place of the money.

Doe any of this ring a bell still being part of our modern wise society. Bloody hell yes, it is our modern society. When Julius Caesar learned of gold mines in Europe he attacked the countries that had the gold, labelled the Kings and people of those countries half-witted barbarians that carried the torch of Satan and was enemy number one of the state of Rome. Therefore to create enemies of the state that could offer gold as ransom and prevent the Romans of paying taxes while the Romans could party in the Arena made Julius Caesar a very memorable man. In fact it made him so famous we know about him as if he was a modern hero. We know him so well because he killed people by the hundred of thousands. We think of him in great respect because he robbed and plundered those that had mines and had money and while he ransacked their villages and raped their woman while slaughtering the men he goes down as one of the greatest leaders in history.

While being one of the most blatant criminals, one of the most outrageous robbers, one of the biggest plunderers of all time we revere Julius Caesar and all others just like him with awe and admirations and we stand breathless thinking about their greatness. They are remembered in fondness because they could rule and create riches. That is a bloody lie. They stripped mimes and minerals that belonged to other nations and murdered them to get their hands on the loot. If they did not plunder in excess we would never have heard of them, and still we remember their greatness in fondness. There is no such a thing as money and money can only be if it is gains as a result of being stolen and murdered for and received by applying the most brutal crime. Luckily that was then and the last such hooligan was napoleon that went through Europe in the name of peace and plundered everything he saw in the remembrance of bringing in civilisation to those that had none.

Remember George Bush senior and the nineties and his election promise of not raising taxes when he did just that just after the elections. Remember the words "read my lips" and he still backtracked on his promises and got the voters raging mad? When George Bush saw he was in a pickle he turned a friend and a partner in war into a household name standing for everything evil Saddam Hussein warred the Iranians with weapons the Americans supplied and bombed the holy shit out of the Persians on behalf of Israel and America. Then just as Bush realised he made a mistake that was going to cost him his second election he found a foe worthy of war and to top the lot this country had oil in huge supply. Then Bush bombed the shit out of Iraq and lots of reasons showed they had no necessity for a reason while they got the war machine going that stimulated the economy.

I can go on and explain how every leader is a war criminal and how they went in to kill Muammar Gaddafi and by bombing cities to bits destroy Libya just to get into a position where the lot had to be paid to "rebuild" the ruined cities and get money for the rubble they brought about in the country that they shot to pieces. The worst crime of all is that money does not exist. All the murder and the plunder are to create something that never was. There is no such a thing as money. If you shout "Gold!" so what about "Gold!"? I ask you where is the hundreds of tons of gold that the Romans went on to kill almost the entire population of Europe? What happened to all the gold that filled the coffers of empires in antiquity? Where has to gold gone to that Alexander the great retrieved after he defeated just about the known Universe during his lifetime? What happened to the hundreds of tons of gold that Genghis Kahn sacked his world he lived in? Where is the gold that the Spaniards brought back from the New World and that was not that long ago? Spain in its present state is economically almost a failing state and yet it brought home riches in the amount of silver and gold by the shiploads. So what happened to that gold because it is no longer in circulation? We don't see the billions that came to Spain or the Netherlands and these countries floated on wealth not that long ago!

Moreover where is all the gold that Britain stole when they murdered a third of my people in concentration camps just over a hundred years ago? Where is the diamonds they sacked from my country when they raped and plundered my people in the very same manner as the Romans did and the Mongols did and the Huns did and Napoleon did not to mention all those I did not mention? Where is the wealth that was created by theft through out the ages? It is gold and it is silver, which is a metal that can't be eaten and can't disappear into mist.

I wish to show you the reader what money is and you can try it in your life. When you have something tangible and solid and you wish to sell it that would mean you wish to exchange it fore currency. In this instance I am not putting you in the shoes of a trader or a merchant because I have explained that part. I am talking about a situation that will come up in the Every-Joes life as he or she goes through life. I will take my selling of my farm as an example. I had to sell my farm because at that stage I had a better chance of dying from a heart condition that I had surviving the heart attack that I was waiting for knowing it had to come. I could not leave my family in the position in which I was knowing they had no chance after my death to get anything of value from a sale as part of my estate with tax and all that being part of the reapers harvest.

The day I sold my farm I told the person purchasing it he had a deal that would never repeat again. The moment I opened my "hand" to accept his money I lost about a third of what I had before the transaction concluded. The moment I accept money and I wish immediately the next moment to return the money for what I had before the transaction the buyer will never accept the money that he paid for the commodity. The moment you take the money you lose about a third of what you had before the transaction concluded. If you wish to buy back what you had you will have to pay about a third more to make the new owner willing to part with what he just purchased. You may put the money down on a new house or a better farm or whatever but then you only pays a certain part of what you had. You then will have to work for many years to gain what you have lost and you may in the end have more than what you had but you will work for decades the be better off than what you were before the transaction ended with the exchange of money. The second you accept money for anything you have of value you lost money that very second, that is if you did not swindle and falsify the condition of what you sell to get the person to pay more for what you have. But when you do that you get on the side where the devil or Satan is and that is the side of the person without scruples. Then you become a Mammonite.

In the present time things are looking up financially all over the world because humans found a way to plunder the earth from energy it stored millions of years ago and had the sun in storage by the way of carbon fossil plants that went to become coal and tar and oil. It is this plunder of fossil wealth that puts the world in such a fine place in which it now seems to be. They take minerals and metals from the earth as if there is no end to the volume in storage on earth and their greed makes them mine as if it will last another ten thousand years…and that is to gain money that disappear as soon as it is mentioned. Yet things are as good as it never was before this time and things can only get better. Man is so clever there is no stop to the ability man may have as long as man does not run out of money. We are a clever lot that can discover anything and create what we can only dream about.

They feed conspiracies to avoid the detection of a true conspiracy. They allow UFO hunters to work relentless because it diverts the attention from the true conspiracy they wish to hide. I personally think this argument in essence is not about "we" or "us" being on the moon or not being on the moon. The make this going to the moon or not very personal by addressing it as "we" and "us" and I can assure everyone I have never been to the moon. But if "we" were or were not on the moon it is something that affects "me" on a personal level. These ideas are fed to all the doubters in order to keep their silly minds occupied and their tongues clicking by those that are "trying" to "stop" the whispers because by having this huge debate over insignificant bullshit is taking every eye off the ball and it diverts much attention from the true issues that science tries to hide. Keep every person wondering about inconsequential nothings and you prevent those investigating to discover the real things that science has to hide at all costs.

This format reminds me a lot about the American Presidential election when every candidate was arguing everything except what the real issues were. It had every candidate fight about personal issues that had no implication about what mattered to the American economy. Bush was fighting some liberal U.S.

Senator called John F. Kerry about that liberal being a true blue blooded Vietnam war hero and that opponent of Bush was fighting to prove that he was fancifully decorated during the Vietnam war and this went on for months while no real issues came into play.

The Bankers paid both to fight the campaign this way in order to keep Bush as President because Bush brought them a war in the Middle East where the Bankers got their hands on oil that brought in revenue unheard of in profits. This is to get other issues debated in order to prevent real issues from coming into focus. John Kerry was paid to destroy the campaign of the Democrats with the blessing of the Democrats in order to give Bush four more years in office as a thank you for the war and all the oil that came attached with the lucrative war. This is how those in charge of our lives run campaigns and how we conduct our lives. They put a lot of trivial bullshit with miner consequence as a discussing topic and then divert all attention from issues that truly matter.

Do you fully believe in science as if everything about science is proven fact and is truthful, never questioning or rethinking one question in having a minutes doubt just because this storey has been repeated for centuries? Are you so confident in science up to the point you will put your life on the line to prove the accuracy of trusting the facts we have in physics? Are you one hundred percent sure about the honesty of science and are you sure about the trust we put in the honesty with which we regard science. This is where the problem starts because science is about money more than the truth. Newton was an alchemist. Alchemist had one goal and that was to find a way to turn lead into gold. What does this mean? It means Newton and all other alchemists wanted to fine a way to cheat nature by changing what is worthless to a product wars are fought over. It was a scheme of finding a way to get rich.

…And there is the problem we have with money. Man can create nothing and that is all. Newton created nothing and Newtonians put it into outer space in such quantities if is overflowing forming the expanding Universe. Money is the only true thing that man created and that is the proof that man can create nothing, not even intelligence! Man thinks of his position as sublimely intelligent and crafty, which is everything that man is not. The only thing that puts us in a better position than our ancestors were in the past is we found a way to harvest energy that was stored from the time life started. It is the sun that sends us heat and it was locked into life and the life turned to carbon and now we burn the carbon to get heat that was heat that came from the sun over the past few millions of years. If it took the earth that long to sore this energy and we waste it in say one hundred and fifty years we are heading for lots of problems which is surviving as humans in time to come. Everything is plundered so that a few fools can feel how it feels to be as incredibly rich as no one ever felt before and then they die and all that wealth is lost. Just like the gold that was bought into money in the past this money exchanged for oil and coal and gas will vanish as soon as it comes into circulation and it will stand to benefit nobody in the very near future because it is money and money is the creation of man and man can only create nothing!

In the society we now have I can only see idiots and more idiots. Look who are our heroes, Pop idols that can sing, if you call it singing? We have movie stars that are gone in sixty seconds as soon as they displease the money moguls that promote movies and financially harvest money. They are created and they are the ones everyone looks up to. They are adored and they are made to please as long as they please. They fill the printing press because the printing press publishes what the idiots with no minds wish to read. You open a magazine and you see some dumb-eyed blond that I accept must be a movie star or a singer or some nut living a glamorous life and she parades in a swanky dress that she will never where again. So the money paid for the dress is gone and her celebrity is gone as soon as the film becomes yesterday's news. However this is not the lot that carries our society into the future. They are not forming the absolute pillars of our community…no we have bigger fools with even much less brains carrying that torch and filling all the promise the future of man might hold.

If you wish to be the pillar of the human intellect you must be able to kick a ball…or hit a ball…or throw a ball… or do something with a ball! On those that carry fame rests the ability to do whatever is funny while doing it with a ball or something that can replace a ball such as a pug or whatever. If you can kick a ball you are paid in tens if not hundreds of millions and woman throw their virginity away after you and men sing your praise and families carry your name or the name on their chests or the brand logo of your team on their shirts just to show all other persons they wish to be associated with someone with your capabilities. What a lot of crap and that hold the intelligence of the human race in ransom.

That is the mindset that will carry intellect into the future. God will help me by getting me mortal and dead because I know by that time my age if not my poor health will see to it that I am dead and I will not be part of what is coming in one hundred years from now. Boy is there sorrow coming when this that everyone believes will never end is going to collapse into a mountain of stinking hog shit. If you can do something that no other can do with a ball you are revered, as a super human being although you are so stupid you can't talk properly or write you name in a sentence or even understand the most basic mathematical equations. Still if you can do special things with a ball that others can't copy you are thought of as holding company with Jesus Christ amongst other big names and you can take the place of God Almighty whenever you choose. This lot with that has a substandard mentality and that has no understanding of facts and are blessed with such meagre wits are the ones on which the society has to build to secure the next and the flowing and the generation after that. By that time we are t of oil and out of coal and out of food and out of intellect. This idiot that thinks a soccer or rugby player is special is the salt of the earth and that intellect and understanding of expectations will secure intelligence, which will maintain civil order in time to come!

I know you are asking what is all this talking about money got to do with science and this is what it holds connection to. I can much more believe in a ghost than what I can believe in money. At least I can create a believable ghost in my imagination but in my imagination I will never create money. Money is a reality that can never be while a ghost might be a reality if I wish to make a ghost a reality. We all and that include me live for money and especially me that has no money. I think of and I think about money night and day because my only fear I have is to be without money because I am constantly out of money all of the time. If I want a cold drink I have to ask my wife. That is how desperate I am for money and yet I hate money because I know money is the root of all evil and money is the force we call the devil or Satan. I am placing this letter in a quest to find money to sell my books to be able to earn a living so I realise what damnation it is to be without money. Everyone tries to prostitute everything for money without showing scruples about it. No one shows a sign of having a conscience because there is an all-conquering lust for money and this has no limits or boundaries. Everyone out there tries to make money by selling improvable suggestions or toxic rumours filled with ridiculous defamation and defamation it is because they can't prove what they declare is true. These rumours could only be fuelled if the Mammonites cashed in on the process by allowing it. Money is the Anti Christ everyone is waiting for. Still try and go without money and you will find how little pleasure life does offer when you have nothing to pay with for any pleasure you want.

We all pray for money as much as we fight for money and we are all slaves of money for what money represents while money is one thing that has no value and is completely fabricated by swindlers cheating the human population into direct slavery. For every job Mammonites create another slave is put to the plough to enrich the greed-grabbing slave drivers that want everyone to see them as being these pleasing-everybody, well-to-do and so kind-hearted money-monsters. Money and what is connected to it is a phantom some make a reality to enslave those that have no money and get us to work for those that has money while there has never been money. From forming a society we were engaged in believing something that can never be. We have to start to give the anti-dote, the antibiotics somewhere to start to get a pure and honest society and not a rich society. What better pace to start with that to get honest in science where everyone believes science is the purest form of honesty. Lets clean up science and the rest will come naturally I believe because then we will go on a hunt not for money but for the truth.

Let's follow the dots and find how this connect with the dots and see where this connects to science and in particular basic physics. We live for and we live by a commodity that does not exist and never had a value and yet in thousands of years millions if not billions of lives went lost to gain money and the power coming from money. The First World War gave up twenty million lives for a fight about money and who ever won the final battle got land and riches from the loser. It was about the one side getting rich to the loss of the other. This is what all wars are about. The Second World War saw fifty thousand lives go lost and if we as humans are prepared to lose our lives for something that does not exist, how the better will we believe in science that does not exist. We except what we are told just because we are told and if we question that which we are told we are true.

In the next pages the tone is getting more complex but it is not difficult because it is I that introduce these ideas and I am just another ordinary fellow with ordinary abilities. What does make it complex is that the reading requires much concentration because it is new ideas that you are going to engage with for the first time ever. Because it is new it requires much more contemplating than what is normally required when reading things that you have heard a million times before. We are brainwashed to except that the only undoubted truth we can ever come across is science. That is where the fable starts because in ninety nine percent of the cases what you read as science is the opinion of someone that considers his opinion to be factual because that person sees himself filling the centre of the Universe. It is just another opinion and most times those that call then scientists inform you about their opinion without showing their case studies so that you can use your intellect to form an opinion on the matter.

If science cannot prove God's existence, it is not God that does not exist, but it is science failing and therefore it is then that specific view about science that should be re-examined since it is the view on science that is proving as being incorrect. This fact is what the so very brilliant and intellectually mindful Newtonian atheist should remember when they fail in their science altogether. That their science fails altogether and that failing it does in all its splendour, is facts I am delighted to prove! The fact is Newton's views were never tested and that the Newtonian views on science were never challenged before and because of that Newton principles never withstood diligent scrutiny before. When Sir Isaac Newton is investigated even in the flimsiest of manners, well accepted facts seems to become very suspect, to say the least. This becomes evident when concluding all the facts this book presents. Now, in this book, for the first time Newton is tested and such testing is the proof you gain by reading that which I uncover. What I bring into the open is unseen facts, which I present you with as I take you on a tour through an avenue of facts I introduce in this work. The lack there is in sensibility concerning Sir Isaac Newton's principles this book proves. The theories of Sir Isaac Newton require proof, which was never given while God never needs proof and that is what science constantly seeks. When science perpetually ignored my concerned calling on and ignored my calling on them because (I suppose) they were finding my concerns wanting, in my final letter to them I promised them never to contact them personally again by any and by all means. I also promised them a fight. This is the fight I promised. I was not worth noticing so I was ignored. I now am calling on the public, as I am ignoring their reputations. I am showing the public just how extremely bright the Newtonian inspired super-thinkers are!

Scientists portray the image that they know all there is to know about everything man might ever know and nobody can ever know more than what they currently know. They tell you what to eat, what to drink, what to think and how to live because they have this image that mentally they are on par with God. They are the utmost superiors on all levels of what forms creation. This attitude applies to Rocket scientists as much as it applies to medical doctors as much as it applies to lawmen. This comes through in all departments albeit language, art, science, law or whatever you may have. If you smoke cigarettes then you will die young because the medical profession found that it is harmful to you while a hundred years ago people smoked ten times what they smoke today and they lived just as long as they live today. But the medical profession took it on them to act as God and force-feed everyone to do what they say or die. The truth is the oxygen you can't live without is burning you to death by aging you and without that you die. If you don't breathe oxygen you die and if you breathe oxygen you die because you are born to die! In the end oxygen kills more people than any bad habit or disease because it slowly kills everything with life. This acting like God envelops all forms of science. What is the truth? Do we hear the entire truth? The truth is that science only reveals some portion of what they know and ignores what is there that they know they don't know. They only reveal what suits their position and never divulge what they know but what does not compliment their view. When science confronts religion they have the opinion that what is in science is everything there is and there can never be more than what science knows or what science wishes to reveal. Science now knows everything knowledgeable and whatever will be known they know. If some scientists are of the opinion that we will fry in boiling water in the next century then it is the Biblical truth because science holds the opinion. Science knows best! Today we laugh at medical practising of a century or two ago and in another century we know the future generations will laugh their heads off when listening to what the informed opinions are of the professionals today. Science has forever veneered their status with this blanket of "they know all". This makes a mockery of the truth because science has no clue why man die or why man age and yet they promise eternal life within the next few decades to come. Ask scientists what is life and they will have an "informed opinion" because they know everything there is.

Science keep up this front that they know everything there is to know while even reading my books prove how little they know about science. They withhold every aspect in science that they do not know and only elaborate in detail that which they think they know about and what we presume they know. They are of opinion that there can be no other way that creation started but according to their science. If the Bible describes how events unfold it then are incorrect because science knows everything. In *The Veracity of Gravity* I show scientifically by using science how creation started precisely as the Bible says word for word but then I also show how little science currently knows about science. The book not in print yet is *An* _open letter Addressing Gravity's Formula_, which is far more elaborate on the matter of how creation started where I show how science proves the Bible correct. How shocking this might be it is just as true.

Science can never take the blame for not knowing. Never is a suggestion put forward that it might be science that holds the shortfall and it is because of science not being adequate that science cannot match the Bible. I can prove how the start came about because I decoded gravity and I did that by finding an explanation about the four cosmic principles. By deciphering the Roche limit, the Lagrangian points, the Titius Bode law and the Coanda effect I am able to show how the very first instant happened when the Universe started the very first point ever formed. These principles are in place and not the principles Newton fabricated... That this book shows. It shows that the cosmos uses other principles than what the Newtonian science promotes. What science says nature uses is not in place or does not hold evidence while what nature does use science deny by just never pressing the issue. I show what is in place and I show why it is in place but first I have to reject what science says is in place because it is not in place.
I show how Newtonians fabricate Newton's ideas about gravity. This is ongoing since the end of the dark ages and Newton. There is no mass that can pull. Most people reading this and who are schooled in physics never heard of the Roche limit, the Lagrangian points, the Titius Bode law and the Coanda effect and these principles are what builds the Universe while I am going to show that there is no factor such as mass. While it serves their purpose notwithstanding never finding evidence to the fact, still science uses _only_ and _exclusively_ Newton's idea of mass while the principles in place the Roche limit, the Lagrangian points, the Titius Bode law and the Coanda effect are never ever mentioned. They sometimes put referring to these principles as law in brackets to deny the status that any of the above law have.

I am going to show you within the next few pages the silliness Newtonian principles hold. While I discuss the principles please see where I am incorrect or going wrong and convince yourself whom is wrong. This is because the Roche limit, the Lagrangian points, the Titius Bode law and the Coanda effect disputes Newton and science would rather discard what the Universe uses than to put a question mark behind the fabrication Newton put in place. Where everyone knows the fabricated information and hiding the reality, which is in place within the cosmos, and that is the conspiracy I show to all. Science stupidity ensures they don't understand the working principles that are in place and that was known for centuries in some cases as the Roche limit, the Lagrangian points, the Titius Bode law and the Coanda effect and therefore not knowing how the principles should be interpreted they hide the concept due to not want to be seen as the ignorant fools knowing the cosmos implements the principles as reality. Science hides their limitations and incompetence behind providing the public selective of information. Take for instance the edge of the Universe they talk so much about. There is no edge of the Universe because there is only an unlimited everlasting Universe out there. What the limits are that they see as the edge of the Universe is the limitation of their equipment that can't trace time back beyond what they see and that serves as their limit in understanding what the Universe offers and how the Universe unfolds.

Trying always be perceives as matching the likes of God science can't face the fact that they can't precede further into space by reading time than what their limitations and their equipment handicap them with, then they put their shortcomings onto the Universe having limits so that they can present the image of total superiority in contrast to limiting the Universe. If they do not understand the four principles in the cosmos and which the cosmos uses as building blocks how can they understand how the cosmos works?

An Ongoing Science Conspiracy Scam

I have this problem... if I say too little I am banished as a story teller without proving anything and if I prove everything I say nobody on earth up to now understood any of my concepts. To explain simple is to be pushed aside and to explain while proving what I say is never to be understood. I am challenging the biggest brainpowers currently on earth and when I say too little I am banished as an attention seeker or I can comply fully by proving what I say and nobody understands a word I say and so nobody reads even a paragraph of my work. I am literally between hell and high water where I am burned as a instigator and a trouble maker, some sort of a joke (and I have written proof of this claim) or when I drown the reader with facts nobody attempts to touch my work because I overburden readers with too many facts (and again I have the letter of a professor putting this in writing). If I ever give up, it would not be because of a lack of insight or because anything I say is wrong but it will be by not finding a method to convey my message.
I have this other problem...I am trying to convey facts about the most difficult subject known to man. It is about Astrophysics. Astrophysics is the science by which we understand how the cosmos layout works when approaching it with physics. The word physics alone scare the pants off everyone and try to match

that fear with the topic that is bewildering then you get people run and hide underneath their beds and shout in fear about persons raping their mental abilities when trying to convey facts in an argument nobody understands. However if you want to understand politics or mathematics or religion or science you have to understand cosmology. Cosmology is the science telling everybody about how the cosmos (cosmology) works and this diverts from Astrophysics where Astrophysics is a mathematical approach to what is out there and cosmology is explaining what is out there. At this point science is very good with astrophysics because they like to tell God Almighty what they think He has to put in the Universe in accordance with their calculations and cosmology they have no idea what is there to explain because what is out there does not match what they tell God Almighty to have that must form the Universe. About astrophysics they are clued up because by inventing more mathematical formulas they create more useless astrophysics that prove nothing and when implementing the formulas they say to God Almighty what He should do in the Universe and they don't look for what is in the Universe that God Almighty put there. In cosmology they don't understand the Titius Bode law, the Roche limit, the Lagrangian points or the Coanda effect or in fact the basic principles that secure the operation of gravity. They see stars don't collide and has no idea why the stars can't collide because putting Newton in reality it must be the only things stars do with ease. They see a galaxy is saucer shaped and has no idea why, they see stars explode and declare the gravity in the star has gone mad, they see most of space is dark and because they know nothing they put that nothing in the place of outer space and in that while understanding no law applying in cosmology they have very little to share except a lot of senseless mathematical fairytale garbage that is worth nothing. However, by using the senseless mathematics they want to teach God.

What you can see is out there and what they say is out there is a mismatch far greater than any opposing ideologies thought up by man. Because nature is always in contrast with their mathematics they hide what is out there and they convey only what they think should be out there not in accordance with what is out there but what should be out there if their calculations were correct. But then they go one step further...they confirm that what is out there is complimenting their views in accordance with Newton. They pretend what they say is what is out there and they never come clean with what they misrepresent because in their view it is God Almighty that is at error making the cosmos different from their ideas.

Everything cosmology stand for and astrophysics try to prove was concept before the streets in London had lights. At the time they concluded what now is accepted mainstream views medical doctors were performing blood letting as the main medical cure, which was done when all else failed and the cure was desperately needed because resulting death was eminent and the only definite treatment was to let bad blood out so that the body could replenish the blood and the new blood would cleans and cure the illness. It was still thought that persons used their hearts to think and the purpose of organs was still to become fully understood. Then Newton devised his ideas but at the time the known Universe that they included as the realm of what is known was that just about only the solar system formed a Universe and knowledge and insight did not even go all the way to the end of our solar system. The last planets were still unknown. What they saw looking through their lenses at the night sky or through the spying glass was what they could see and was the entirety of Universe and that was very little. They did not even know about all the planets we know of today. The idea of comets returning cyclic was a thought still to become a reality.

Then Newton gave us mass. By having the mass one would find the force that drives the moon towards the earth. Before the event of Newton's miraculous discovery of forces driving planets around the sun to the tune of gravity, this was mainly God's prerogative to have such knowledge. No one gave a fart about understanding anything this formula said because if they did they would know the correctness thereof was a joke. So, with nobody having the intellect to argue the correctness thereof at the time it was introduced, it was considered brilliant just to pretend one do understand and then become a member of the Brainy Bunch society ruling physics. You only have to make believe and pretend what nobody can ever understand to get so wise that you can assemble your own Universe by applying mass. All one needs is to pretend to understand and believe that mass pulls mass and then you are as clever as Newton and everyone knows Newton is bloody clever so anyone believing the Universe is contracting by the measure of mass that forms gravity places such a person equal to God's intellect. All you have to be is to be stupid enough not to ask silly questions and then you are brilliant. Be a Newtonian and follow your leader blindly with no questions asked about how gravity applies mass. Everyone would be so amazed with your brilliance they will think of you as one of the Brainy Bunch. To others you say when asked to explain what gravity is that they are intellectually inferior when they don't understand Newton. People will stare at you in awe and admiration and all you have to do is to believe Newton and believe in science. Now Hubble went along and burst this friendly little bubble by his unasked for discovery of the expanding

Universe. How easy could it be to be so smart you think of your position as equal to God's position? No one dares to argue with you because you have Newton backing you. Only the insanely stupid or mentally handicapped will dare to argue with Newton about mass having pulling forces going around as gravity.

That way all that anyone is asked for when to be thought of as being a clever human is to repeat after Newton that $F = G \dfrac{M_1 M_2}{r^2}$ is the formula on which the entire Universe stands by the principle of physics.

Newton never proved that $F = G \dfrac{M_1 M_2}{r^2}$ was correct but since nobody understood whether it was correct or not nobody gave a blue apes virtue about the concept of correctness. Repeat one million times mass pulls mass until you believe it and then you are a physicist. This illusion was ending and it came with a bang that broke a two hundred year conspiracy, which guaranteed silence about scientific misconduct.

Newton as person did believe in the art of black magic and Newton was a certified alchemist, a fact nobody dwells on much. That is why he got this force that pulled by mass and his force had a spiritual reality according to his alchemist background. He saw the force pull everything towards everything and that is how everything will end. When everything collides with everything the pulling must stop because then everything united with everything else. Remember at the time it was still uncertain if the world was truly flat because after Newton many tests were conducted to this effect and they formulated absolute proof long afterwards. Still to have the formula $F = G \dfrac{M_1 M_2}{r^2}$ put you on par with God because if you knew the mass you could design you own personal Universe just to your liking. You had to know the mass and then you could calculate the force and then you knew what God knew so in that sense science became as clever as God is. That prehistoric backward ideas and view are still professed today. Medicine corrected their mistakes. Every part of science progressed except for Newtonian views, which remained as backwards as it was centuries ago but is still enforced on students even today. They cling onto this knowledge about mass that removed the essence of having a God because now physicists could become as clever as God and replace God just because mass is pulling whatever together. Then came E.P. Hubble!! Have you heard about E. P. Hubble? He is the man the Hubble telescope is named after.

Let me inform you about Edwin Hubble the man that (almost) derailed Newton. At the time Newton formulated his impression of what is the correct cosmic ideology that is still accepted to this day it was thought Africa had dragons and ships sailed because wind power was the ultimate source of energy. They were dabbling with steam but the development of boilers still had room for much improvement. Canals were new innovations and boats were horse drawn. They just began to understand lightning and police was a term that stood to become invented. The information was desperately poor but it portrayed physicists as equal to God and that was enough. Physicists could now claim the honour that which all humans in the past bestowed on God. This they still cling to because they still cover for Newton and it is because of this that they can walk on earth pretending they know what God should have known at the time God created a Universe…except now they don't need God any more because now physicists turn out to be good enough to have around. The idea of being and acting God and rule in the way God can suit them fine. To protect their image of being supremely clever they had to save Newton to save them.

There was one culprit that was responsible for this information that Newton was wrong and had a chance of leaking out by getting everyday news and he was E.P. Hubble. Before Hubble got so outspoken about his findings the entire world of physics new they were bullshitting the public blind about Newton's anomalies but was getting away with it for centuries. Newton said the apple fall because it had weight but then they changed it to mass. Everyone on earth got confused between what is weight and what is mass. Everything standing on earth holds weight but however you use weight when measuring mass, mass has nothing to do with weight. Mass contracts and weight weighs. If you have weight which all can see you have mass but mass contracts and although using the measurement weight uses mass pulls in place of weight weighing. Planets spin around the sun because they have mass and because they have mass they spin in an ellipse. How the ellipse comes to form I have no idea because the mass doesn't change anywhere. Just go and question "Kepler's" laws that Newton founded and you can see how far Newton was off the track. Everything that didn't have weight had mass and mass was not the same as weight. Therefore everything contracted by the measure of mass but mass should not be confused with weight. E.P. Hubble saw through his telescope that the Universe was moving apart. Before this everybody agreed with Newton that the Universe was contracting under the load of mass. Newton saw the moon coming closer by the value of mass as the earth was getting closer to the sun by the value of mass and to be equal to God those intellectuals could just apply $F = G \dfrac{M_1 M_2}{r^2}$ and redesign the entire Universe according to each ones liking. If you had the mass you could gauge when the Universe began. If you had the mass you could gauge when the Universe will end. By having the mass one would find the force that drove the moon towards the earth. Before the event of Newton's miraculous discovery of forces driving planets around the sun to the tune of gravity, this was mainly God's prerogative to have such knowledge.

I wish to spend some time with this notion. Hubble saw the Universe was expanding. Everyone looking through the telescope saw the space was growing like bread rising in a hot oven. The first thought would be "oh gosh, Newton had it wrong" because the Universe is expanding and not shrinking as Newton said it does! The Universe is going the other way, which makes Newton look bad. We have to dismiss Newton and start to study what is going on in order to see where we are and for that matter where is Newton wrong! This will come to a new field of study but first let's get rid of everything Newton said because mass is not pulling mass and the cosmos is not shrinking. Did they come to realise Newton had it wrong...no they had to vindicate Newton. They had to find the missing mass that the Universe is being short on. It was clear at any given instant the Universe was expanding. It was moving in a direction Newton did not foresee and whatever was decided did not change the direction of movement in the cosmos. The blame for the mistake went to the cosmos for not doing what Newton said the cosmos was doing.

Then along came a man that had a good look at the Universe. The discoverer that researched the Universe night after night had a name and a position of seniority, which prevented others from pushing his opinion aside and derailing his career. The man was E.P. Hubble. Hubble saw the Universe expanding while Newton said the Universe was contracting. This man Hubble became a fly in the ointment. All Hubble had to do is to tell those not believing him to look through his enormous telescope so keeping Hubble quiet was not easy and to try to discredit Hubble would be most stupid. Through his telescope any one could see that the Universe was expanding and the expansion was most rapid. He looked at the sky and concluded the lot was not shrinking but it was expanding. Any one that would look through his eyepiece could clearly see the lot was not shrinking. The lot was growing apart. In some cases he said the lot was racing apart. The Universe was growing by miles and not shrinking into nothing.

The man E.P. Hubble proved through his telescope that the Universe was expanding and not contracting which made the Universe quite wrong. Yes you see correct...it made the Universe be wrong! Newton could never be wrong because Newton was never wrong yet...so if the Universe is out of step with science, then science will correct such an abnormality by finding a way to defraud science and postpone the correcting that the Universe had to comply with since the Universe owed the Master Newton some apology. Did the Universe not know that he whom never can be wrong is in name Isaac Newton! Decisive action was needed. When will the Universe confirm its incorrectness by affirming Newton's obvious correctness? The smartest brains on earth came to a table to decide how could the Universe oblige to once again adhere to Newton and start a contracting and abandon this rebellious expanding. It was not Newton that had to comply because if Newton was seen as being wrong then all physicists had to admit they had everything wrong about the Universe and they were not intellectually on par with God Almighty. No, God Almighty is wrong and had to conform to apply to Newton's ideas so that physicists could remain equal to God Almighty. The question the clever ones had to answer was when will the cosmos come clean and prove Newton correct. God made this mistake and it was the duty of physicists to find how God could apologise to Newton and correct the wrongful manner that the Universe went about. They had to find enough mass to get the Universe to make a U-turn and start the contracting and this had to come as quickly as possible. E.P. Hubble saw through his telescope that the Universe was moving apart.

This man Hubble with his large telescope wrecked everything science was based on and never even flinched about it. He almost brought physics to a standstill. Everything was going well and everybody was enjoying being clever as long as you were stupid enough to be a Newtonian. Then this man unveiled a fact that declassified Newton as a flawed worthless form of stupidity. The Universe expands and does not contract. Now once again only God was flawless with Newton flawed throughout. This the Brainy Bunch was not going to appreciate and this did not go down well with the most esteemed of the Newtonians. They were not going to give up their ability to redesign the Universe and be God that easily.

They all came together and decided not to go in search of what is correct, no they decided to look for missing mass because the cosmos was hiding mass they did not see. Only if the cosmos cheated and misplaced mass somewhere could this be true. They had to find what game the Universe was playing to try to cheat Newton. Look at the picture and see the density there is. But then in some large areas there is no density and only darkness and they put the darkness down to nothing being there so in those regions the density was nothing and in other areas the density was overwhelming but nowhere was the density evenly spread to give a even value that can apply over an area say the quarter of the Universe.

This unleashed a problem the world had no name for. Everything known to science was at that point devastatingly unknown to science. The world was expanding and not contracting which made the Universe quite wrong. It is impossible to have any vision about Newton being wrong. Newton could never be wrong because Newton was never wrong yet...so if the Universe is out of step with science, then science will correct such an abnormality by finding a way to defraud science and postpone the correcting that the Universe had to comply with since the Universe owed the Master Newton some apology. Did the Universe not know that he whom never can be wrong is in name Isaac Newton! Decisive action was needed. At this point I cannot believe that the most brilliant minds were so naïve and therefore I must suspect deliberate deception. Hubble was far too prominent to blow away and Newton was found wanting. At that point they put the onus of proof not on Newton but turned the focus away from Newton to what the presented as the guilty party. When will the Universe confirm its incorrectness by affirming Newton's obvious correctness? If they had to admit that Newton was wrong, the most intellectual science then had to admit they had nothing to show for all their minds brilliant work.

Science that was defying the likeliness of a living God stood bare and naked for all to see. They put the onus of proof and converting onto the cosmos. They asked not Newton but the cosmos when will the cosmos come clean and prove Newton correct, maintaining their unshakable belief that even the cosmos could be at blame but Newton could never be wrong. . When will the cosmos admit to a mistake and set its crooked ways straight. When will it meet its diverting from Newton and reach a point where the Universe will finally come to comply with what Newton demands. It is the cosmos that is wrong therefore it is time to find out when the cosmos will correct its manner.

In order to cover up for Newton's misperception an entire variety of reasons are established, each accepted as a possible truth. The fact that Newton's principle goes begging never gets mentioned, although the only reason why it would never get mentioned is because it is the only valid conclusion and that they don't want. All other reasons they mention is overruled by Newton's principal of mass pulling. Mass pulling is the founding law that all other factors rest on. The earth slinging the Moon away can't be a factor because the mass of the earth is too great. The mass that pulls reduces the radius by the square.

This unleashed a problem the world had no name for. I blew physics apart. It cracked what was most solid before. Hubble proved everything is expanding in contrast of general opinion about a contracting Universe. Everything known to science became at that point devastatingly unknown to science. The Universe was expanding and not contracting which made the Universe quite wrong. The blame had to go to the Universe. It is impossible to have any vision about Newton being wrong. Newton could never be wrong because Newton was never wrong ...so if the Universe is out of step with Newtonian science, then Newtonian science will correct such an abnormality by finding a way to defraud science and postpone the correcting that the Universe had to comply with since the Universe owed the Master Newton some apology. The Universe had to carry the blame for the audacity to have insufficient quantities of mass.

Did the Universe not know that he whom never can be wrong has the name Isaac Newton! Decisive action was needed. At this point I cannot believe that the most brilliant minds were so naïve and therefore I must suspect deliberate deception. Hubble was far too prominent to blow away and Newton was found wanting. At that point they put the onus of proof not on Newton but turned the focus away from Newton to what they presented as the guilty party. They had to determine when the Universe would confirm its incorrectness by affirming Newton's obvious correctness and start to contract like Newton said and not expand as the rebellious cosmos is doing! If they had to admit that Newton was wrong, the most intellectual science then had to admit they had nothing to show for all their minds brilliant work. Then they would be uncovered by their stupidity being revealed and not be praised as the only wise in the world.

Science that was defying the likeliness of a living God stood bare and naked in disrepute for all to see. They put the onus of proof and converting onto the cosmos. The question was when will the cosmos come clean and prove Newton correct by having enough mass to contract as Newton insisted it must do. When will the cosmos admit to a mistake and set its crooked ways straight. When will the cosmos meet its diverting from Newton and reach a point where the Universe will finally come to comply with the standards that Newton demands. It is the cosmos that is wrong by going about expanding therefore it is time to find out when the cosmos will correct its manner and adhere to the superiority of Newton. To deal with such a task they needed a man with a bigger ego than he had an IQ.

They needed a person that thought more of his abilities than his ability to grasp any complex situation. They needed a man that was presented as a genius without ever proving his genius. They had a man that filled the centre of the Universe, which then placed the man in a location so high in status from where the man could see the entire Universe.

They had just such a man. He went by the name of Albert Einstein.

They devised the critical density theory. How smart can those who are smart get when they truly try to be smart to save their skins? How brainy can the Brainy Bunch be when they deceive a plan to outfox all of Human kind? You can go to the Internet and read all the multiple arguments about the Universe going flat and contraction therefore come about although not one sod amongst them know the least how singularity works.

Here it is in a nutshell; what this is, is not a theory that is called the critical density theory but it is a criminal venture conceived by the intellectual minds in physics to go criminal and defraud the world. They decided that the blame of the Universe not working as Newton said it does needs to be placed at the door of the Universe. The Universe is lacking matter to contract as Newton said it should. Newton still remains absolutely correct. The blame for the mistake is diverted to the cosmos and away from Newton. Newton did not lack the insight to see the Universe expands, no the Universe lacks material to contract.

The Universe made the flaw and now we must find how the Universe will correct its flaw. The Universe went wrong by not having the sufficient mass and therefore the Universe must correct the lack of mass to get the Universe back in line with Newton. The blame for the mistake must be laid in the midst of the material within the Universe. Newton's formula stands correct and all suspicion goes in the way of how the cosmos was designed not to applaud the truthfulness of Newton.

All those unrealistic arguments the Brainy Bunch offer as to why the missing mass or dark matter will bring a clarifying solution to avenge Newton has one damning flaw. Whatever they bring as an argument is tainted by a law in mathematics. It is built into the formula $F = G\dfrac{M_1M_2}{r^2}$. If the radius increases, then the value of the mass reduces while staying the same. If the factor representing the radius r^2 becomes $2\,r^2$, then by the very same token does the mass become half of its previous value! It is effectively this $F = G\dfrac{M_1M_2}{2xr^2}$ is $F = G\dfrac{M_1M_2}{2}$ and this will bring about that while the cosmos is expanding, the worth of the mass is reducing by the same margin. This is not rocket science; this is mathematics at its most basic. If the Universe was expanding then the measured value of the mass was declining that is if mass was responsible for producing contracting gravity. They are the ones that are the masters in mathematics. They are the ones that know mathematics better than anyone else on earth…and they missed this truth. This missing the basics was as deliberate as it was swindle the hide Newton's incompetence and with it their failure to understand physics. This is where the second conspiracy started.

With all this in mind did any one ever come to wonder about the all too famous Einstein's critical density theory and the fact that this idea was conceived to conceal the corruption of Newton in physics? The fact in truth is that the Einstein's critical density theory was a scheme plotted by those in charge to cover up and conceal corruption in the heart of physics. If Einstein was unable to recognise the most basic of mathematical principles then what type of genius did physics create in him and what slur did physics promote. This idea of the two factors being in opposing relevance is so simple that children will recognise the principle, and yet those fathers of physics wants me to believe that the greatest mathematician that ever lived did not realise this principle…the principle that the radius and the mass stands related and the growth in the one will promote the decline in the other as a dominant factor. This puts the critical density idea down as a fart. While the radius grows that puts distance between materials with the cosmos growing the increase in the radius will reduce the potential role that mass plays in the future. Fort all the mass that might come into play the fact that the radius is growing in time dismisses the chance that mass in the future will have any significant part in cosmic development. This basic principle of the critical density prospect is thereby dismissed and that is by pure mathematical equating laws.

However as bright as they might be mathematically they are clueless about mathematics. This is the mathematical truth…as the dividing factor increases; the influence that the mass will project in the formula

$F = G \dfrac{M_1 M_2}{r^2}$ will diminish in respect to the growth of the distance. In that sense the gravity force between the earth and the moon must reduce its ferocity therefore weaken because more distance reduces the value that mass projects. The higher the distance becomes the less will the influence of the mass be when divided by the increased radius. However, Newtonians only apply their ability to calculate and knowledge for the purpose of upholding Newton and never to provoke Newtonian liability by telling the truth. Never is there any mention of mathematical reality when mathematics is used. However this knowledge about misinformation no one ever knew to be true outside of the intimate upper circles of physics, and this unmasking was to be prevented at all cost. A plan was to be devised because if the public found out Newton was a fraud all along and being Newtonian was the personification of stupidity the entire science world would come tumbling down on the heads of those most important Brainy Bunch.

Can any one with this information including the information given on the previous page have any other conclusion? It is obviously clear that having such a total idea that there might be dark unseen mass floating in the Universe which at this time does not generate gravity but will some day because Newton has to be correct at some point in the future. I am to believe that dark undetected mass can be found and such undetectable   mass could be found which will bring about contraction after all this expanding? Why would the mass at present then not activate gravity and why would the mass at some point spring to life and start activating gravity? How much can the Physics paternity still hide the fact that Einstein's critical density is being used as a cover-up to distort the truth to conceal fraud? The uncovering by the Hubble constant about of the Newton fraud is so simple to see. Hubble found the Universe is expanding and Newton's said otherwise. Who is lying about what? Hubble's declaration was on track to blow the cover that was concealing the Newton fraud wide open and uncover the century's old deception. To see this we have only too look at the comet behaviour when any and all comets again come around on a cycle by repeated visiting the sun. The question is if it is mass pulling mass onto mass, then why do we have comets left in the solar system? The mass of the Sun should by now at least have destroyed every comet going around.

Every indication that we so far received in vivid portraying from astronomy photography studies from outer space disputes a shrinking universe concept. From the moon increasing the radius distance between the earth and the sun, to the Hubble Constant indicating a space growing any where in space wherever man may conduct studies. Since the end of the middle ages a force called gravity was identified, but more than that science did not take it. What is gravity, besides being a force? What forces the force? I introduce a cosmic theory that turns the missing questions to answers.

For all the genius Einstein had, Einstein failed to see the most simplistic and tiniest mathematical rule. Einstein failed to realise that if there was insufficient mass at the beginning of the expanding Universe, the growth of the Universe will reduce the influence of such mass as a factor further because as the radius grows, such growth will restrict the gravity by rendering the mass progressively more incompetent. However, it was more important to acknowledge that Einstein filled the centre of the Universe because that is the only place Einstein could be to calculate all the mass he saw that filled a Universe.

If the Universe is expanding as Hubble indicated, the growth of the radius will reduce the influence value of the mass as every second passes. The mass will become more and more wanting for such a task. Yet with this obvious shortsightedness about the most fundamental mathematical principle the mathematical it was this that the genius Einstein failed to acknowledge, Genius Einstein saw him fit enough to calculate and measure something as overwhelming as the Universe. As in the case of Newton, Einstein was an ego driven maniac that saw his abilities fit to measure and master the Universe while his mind was too simple to recognise the most basic principle of mathematics, the principle of relevancies or ratios. If you put something in an equation in division of the top it is the bottom part of the equation that is most important and not the top part.

This is the mathematical truth…as the dividing factor increases; the influence that the mass will project in the formula $F = G\,\dfrac{M_1 M_2}{r^2}$ will diminish in respect to the growth of the distance. In that sense the gravity force between the earth and the moon must reduce its ferocity therefore weaken. However, Newtonians only apply their ability to calculate and knowledge for the purpose of upholding Newton and never to provoke Newtonian liability by telling the truth. Never is there any mention of mathematical reality when mathematics is used. However this no one ever knew to be true outside of the intimate upper circles of physics, and this unmasking was to be prevented at all cost. A plan was to be devised because if the public found out Newton was a fraud all along and being Newtonian was the personification of stupidity the entire science world would come tumbling down on the heads of those most important Brainy Bunch.

No, better still, to save science a conspiracy was devised. The most intellectuals on earth had to cook up something and devise a plan to save their image and the name of Newton. The most intellectual minds concluded and fabricated to form a conspiracy to withhold the truth and forge a fraud that lasted almost ninety years to the day. Should anyone disagree with the term conspiracy, then please let me know what you would call what happened after Hubble's discovery became prominent on the news.

In the past it was accepted that only Newton and God never made a mistake and since the Brainy Bunch Newtonians were mostly atheistic or atheistically orientated they were not that sure about the credibility of God but the unquestionable accuracy of Newton those atheists were pretty sure…and now it seems Newton made a mistake. That just could not be. They would rather have God make the mistake than let it seem as if Newton was up to no good while Newton made the incredible mistake.

Then they allowed God to make the mistake. If the Universe was expanding it was God's fault. It sounds much better than have anyone think it is Newton's mistake. If the cosmos expanded while Newton said it must contract then this rebellious behaviour of the cosmos must end. They had to put the blame on the Universe and then ultimately on God for making such a mistake. It must be God that made the error.

In $F = G\,\dfrac{M_1 M_2}{r^2}$ the size of the radius by the square will determine the outcome of the equation's probability and not the top part holding mass. What a mathematical genius that one turns out to be. While the radius enlarges, at the same proportion does the influence of the mass factor reduce and the mere fact that the radius increase shows that at no stage further into the future can the mass stem the growth of the radius because the radius overpowered the mass factor already. Unless there is new material entering the Universe at a point, which is impossible, the entire concept is fraud.

No, better still, to save science a conspiracy was devised. The most intellectuals on earth had to cook up something and devise a plan to save their image and the name of Newton. The most intellectual minds concluded and fabricated to form a conspiracy to withhold the truth and forge a fraud that lasted almost ninety years to the day. Should anyone disagree with the term conspiracy, then please let me know what you would call what happened after Hubble's discovery became prominent on the news.

Science goes even much further. They sequestered Albert Einstein to measure all the mass in the entire Universe to find out when will the Universe start contracting and come to the end of its life cycle in accordance to Newton's gravitational pulling principles. This action shows that there is no limits or ends to which they will not go to find the end or bring finality to whatever they try to establish. To measure the end of the Universe is going much further than to establish when the earth and the moon will meet their gravitational destiny, with Newton and his theory of mass pulling mass so fundamentally proven. But there is no end to their resolution for they never stop with their inquest. When the Critical Density Investigation did not deliver the results that would bring satisfaction, they went in search of Dark Matter. I am going to go much deeper into the Critical Density and the search for Dark Matter later on in the book.

No outcome of sorts was necessary because they only had to shift the blame of wrongdoing from Newton onto the cosmos and the bluff was on. The idea was never to admit wrongdoing on the part of Newton and Newtonian science but to post pone, delay and divert attention away from the truth. If there was not enough mass to start with, no dark matter can kick in later on and start secondary mass frenzy that at that stage will then be enough to bring about the required mass potential that will turn the Universe around from expanding to contracting.

To establish a scenario that would hide all deception they got the man that has a bigger ego than an IQ, and they tell the world this man is a genius while the fool does not know the least of mathematical principles because his Master Newton did not know the least of mathematical principles and so they got him to measure the Universe. I'd thought it would keep him busy for some while say the next billion years or so but he was cleverer than that! While they did not even have any device (and will never have such a device) through which anyone would be able to see what the entire Universe holds, they set of a scandalous misconception that Einstein could calculate all the mass in the Universe.

Off course as can be expected, there was not enough mass and there will never be enough mass because there is no such a thing as mass in the entire Universe. When the deceit played out to the full, the fraudsters being the paternity of physics elaborated on the delusion by trying to find dark matter that is hidden. If the dark matter did not develop enough contraction by this time, there is no chance in the future to develop enough gravity because the factor of what mass supposedly should have is tarnishing as the Universe expands. The bigger the radius becomes the less would the mass effect be notwithstanding.

With everyone in science saluting Newton's gravitational contracting there was an extended effort by Albert Einstein to find the critical density of the Universe. That is the backbreaking effort that science took with painstaking accuracy to find the density that the Universe must have to start to contract as Newton said it happens. The critical density idea did not pan out and that left science high and dry for answers about science. They did not stop there, no Sir, they conspired more by exploring onwards in darkness to find the answer. The answer had to be in place so that Newton must come out in the end as the only idea that could be correct about mass pulling mass closer.

The idea was never to admit wrongdoing on the part of Newton and Newtonian science but to post pone, delay and divert attention away from the truth. If there was not enough mass to start with, no dark matter can kick in later on and start secondary mass frenzy that at that stage will then be enough to bring about the required mass potential that will turn around the Universe from expanding to contracting. To establish a scenario that would hide all deception they got the man that has a bigger ego than an IQ, they tell the world this man is a genius while the fool does not no the least of mathematical principles because his Master Newton did not no the least of mathematical principles and they got him to measure the Universe. While they did not even have any device (and will never have such a device) through which anyone would be able to see the entire Universe, they set of a scandalous misconception that this Einstein could calculate all the mass in the Universe.

Off course as can be expected, there was not enough mass and there will never be enough mass because there is no such a thing as mass in the entire Universe. When the deceit played out to the full, the fraudsters being the paternity of physics elaborated on the delusion by trying to find dark matter that is hidden. If the dark matter did not develop enough contraction at this time, there is no chance in the future to develop enough gravity because the factor of what mass supposedly should have is tarnishing and tarnishing as the Universe expand. The bigger the radius becomes the less would the mass effect be.

The community of astrophysics are trying to frame a picture where they set the stage in the way that if the Universe were stretched to a point the mass would not tolerate any more expanding. The mass will get frustrated in some way and show resistance to the increasingly elastic expanding. The gravity constant (I suppose) must prevent any further expanding. How they ever got to such an argument I never could tell. They surmise that outer space is consistently overall filed with nothing and when this nothing is stretched to the limit, the nothing would resist in growing more nothing or become further nothing and the nothing would stop other nothing to enter outer space in the community represented by nothing. If ever there is a faculty ruled by absolute inconsistency and rubbish as the motto of logic it has to be astrophysics.

Every measured kilometre represents nothing. Every mm is one of nothing. We on Earth are 149×10^6 kilometres holding nothing away from the Sun. Only they can argue that outer space is nothing with material here and there. If that is the case then which has more nothing between the Sun and Pluto or the Sun and Mercury. The distance between the Sun and Pluto is more, therefore that which outer space is made of is more than in the case of Mercury and the Sun. Therefore Pluto has more nothing between the Sun and the planet than Mercury has between the planet and the Sun. Only astrophysics and all the geniuses guarding the principal of astrophysics can put a calculated value by measure on nothing. In fact Mercury has hundred times less nothing between the planet and the Sun than is the case with Pluto.

Since my days at school I was always under the impression that a hundred times the value of outer space being nothing is numerically expressed as (zero = 0 x 100 = 0), but where the genius that is such a prevailing part of astrophysics take the stage we find that Pluto can have 100 times more nothing than the amount or distance measuring nothing than Mercury has. The figure containing nothing that puts Pluto at the edge of the solar system is one hundred times more nothing than what Mercury has where Mercury becomes the first planet in the solar system. That is astrophysics. The brilliant minds of the mathematicians hold no rules apart from what they can calculate. Astrophysics is the only department throughout the Universe where normal rules don't apply since because with mathematics they can bend all laws as they wish…in fact Newton started the trend with his deceit.

Only the guardians of astrophysics policy can know why the undetected dark matter will start producing gravity to change the expanding to contraction. Would the fact that it is detected, change the influence it established? Or is it merely to extend the cover up and allow the deceit to linger until the following generation. There is no mass and any one that says there is mass, let such a fraudster then explain why all the planets irrespective of size or density, spin around the Sun at the same sped as all the others. Let them prove that the Universe acknowledge big and small and let them show how Jupiter can move at the same pace as does Mercury and Pluto while Jupiter is so many times more massive than the other two mentioned. More condemning evidence is yet to come because the astrophysics tricksters did not leave the corrupting of evidence just at that.

The fatherhood of physics never once diverted from acknowledging that Newton's contraction is the prevailing thesis on which the cosmos is built because they accepted that Newton used unlawful arguments and to cover up Newton's fraud which they still use to this day, they then proceeded with further criminality when producing the bluff they established with Einstein just to fool everyone in the normal public. Without ever recalling Newton's contraction theory that is obviously not working or admitting doubt about Newton's testimony to the effect, physics accepted the Big Bang Theory. The Big Bang theory opposes what ever Newton might have implied. The physics paternity however finds it wise to still advocate Newton while admitting to the Big Bang event. Newton said the lot is contracting. Go on and marry that with the Big Bang that says everything is expanding. You can't promote both except if you have an able argument you can apply so that you can define why we would see the two merge.

According to the Big Bang concept it would be that the Universe comes from a point the size of a Neutron. That makes the radius parting the Universe infinitely small. It just about removes the radius as a factor. At the very same implication it takes the pulling of the mass (if there are pulling forces converted by mass) to a level it will never again have. As soon as the distance between the objects holding mass started to grow, the power and influence of the mass factor started to diminish in the same ratio. If the mass were incapable of contracting the Universe then, it will forever remain contracting the Universe.

Then you may ask what the story is? Read on and you will learn how far Mainstream Physics stray from the truth and how big a cover up the paternity is protecting.

They left no stone untouched to come up with a conclusion…and yet not one person in all that time started to think that the contraction will be much better monitored by researching the moon reducing of the radius it has between it and the earth. Would it not have been much easier to study when the moon will splash into the earth and from there work out when Newton's attraction will have the worst collision we can think of happening in our backyard? No one was after the truth because it was all about vindicating the correctness of Newton. So someone got very smart and even more conniving! Someone invented dark matter. If you can't see dark matter it means you can't not see dark matter either. So if you can't prove dark matter exists no one can prove there is no dark matter in existence. This was brilliant, befitting the thinking power only physicists could achieve.

In the past it was accepted that only Newton and God never made a mistake and since the Brainy Bunch Newtonians were mostly atheistic or atheistically orientated they were not that sure about the credibility of God but the unquestionable about accuracy of Newton those atheists were pretty sure…and now it seems Newton made a mistake. That just could not be. They would rather have God make the mistake than let it seem as if Newton was up to no good while Newton made the incredible mistake.

Then they allowed God to make the mistake. If the Universe was expanding it was God's fault. It sounds much better than have anyone think it is Newton's mistake. If the cosmos expanded while Newton said it must contract then this rebellious behaviour of the cosmos must end. They had to put the blame on the Universe and then ultimately on God for making such a mistake. It must be God that made the error.

The easiest is to put the blame onto God by finding the fault at the door of an uncompromising Universe. It is and must be nature not playing ball with Newton and it must be God that is behind it. God designed the Universe wrong just to be difficult and now Newton is getting the bad name. This had to be corrected and the blame for this had to be diverted back to God because after all Newton can't do wrong so it had to be God that made the blunder! To make the conspiracy believable they had to conjoin a concocted story that would have every halfwit on earth believe it. What will make the Universe expand? It must be a lack of the something that makes the Universe contract. Not enough contracting solution will be the cause of the expanding. If it was mass that had the Universe contract in Newton's terms, then a lack of mass will lead to a shortfall in gravity and then the elasticity would not be enough so the elasticity would be tested and before the elasticity failed completely they had to get the Universe back on track. They saw a good measure of darkness splitting small bits of light. This must be it then. Who put that much darkness amongst that little Light?

Look at the picture and see what is out there for us to see and to digest because what you see can only be the truth and more true that can any form of ideology ever be. No theory can present this factual picture. If it was mass that should contract the Universe then it had to be not enough mass that would be to blame for the expanding of the Universe. Now to get someone credible enough in the eyes of the public yet foolish enough in ego to provide the cover for the conspiracy to work was another matter. The only candidate must be Albert Einstein. They had to get Albert Einstein to measure all the mass in the entire Universe. Now this is where the joke no longer seems funny. Only fools and idiots are going to fall for that. Look at the picture on the previous page and the picture above. This is unrecognisable small portions of a large and overwhelming large Universe. Who is keeping whom for a bloody fool?

Is there anybody that will seriously try to convince me or any other sober-minded person that any human being can measure what is to be considered as mass even in the picture above? Go back to the previous page and look at is presented as material in that picture. Is any sanity left in the suggestion that any body may even think of attempting to calculate what is in such a small portion of a fragment of a sideshow of the Universe?

Can there be any person in his right mind that will think he could have the ability to measure only what is in these pictures, let alone what might be available in the entire Universe? If anything can ever bear testimony of how mad those Newtonians got, then this must represent their total loss of mental coordination about what is reality and what is hallucination of a mind gone missing of reason. Their arrogance at that instant of deciding in following a direction grew into mindless stupidity.

Put the earth in any of these pictures and the task is senseless to perform. Put the sun in as a visible star and the task is ludicrous. Put the solar system in and still it will not show as a freckle. The entire Milky Way might come about as a speck somewhere, but not big enough to be noticed. Then tell me please how many Milky Ways might fit into this small part of a huge Universe…and Albert Einstein was prepared to measure the entire Universe. I know the formula he used but using that formula indicated just how lubricous the attempt was. Still, they say Einstein determined the average mass of the entire Universe.

This is a small part of what he said he could achieve. This is a conspiracy as blatant as ever there was one. Believing this first require the drinking a bottle of rum and then getting high on a barrel of cannabis. Either they were fools or they thought the entire human race was brainless fools incapable of thoughts.

In the Theses the explaining gets a lot more technical but now we have to get back to the conspiracy and show the silliness of those hiding the conspiracy and posing the funny part as reality. The conspiracy and what it represents in information becomes a silly joke…and it works, as a conspiracy because as far as I can trace I am the first not to be fooled by the fantasy of the fable theory underwriting the serious part! All the models and theories show the Universe expanding while every one needs a Universe to contract. This was not helping Newton while in purpose it should be helping Newton.

If everyone saw that Newton was a blubbering fool that was mistaken about the cosmic principles that the Newtonians underwrite as better-than-Evangelic-Gospel then they all were a pack of idiots that new nothing about what they professed that they know everything about. They had to get around this because now all theorists were ganging up to abandon the drowning ship representing Newton and came up with theories how the Universe work by expanding. All the models had one thing in common they didn't represent Newton's contracting principles because the growth supported Hubble by "being bigger", abandoning Newton's "getting smaller" idea.

Then the most powerful minds the human race has to offer set in action another plan. The first plan of finding missing mass did not work because Einstein did not locate enough mass to surpass the critical density level. Moreover he was so fast the gig was over before it started fooling anybody' Then they devised a plan that was beyond detection, a plan only the biggest masterminds in physics could devise. This plan surpassed all other thinking that came previously. They had to keep the image of Newton beyond suspicion so that everyone would think those that were the Masters of physics knew what physics were. Their futures depended on Newton's future. The con was up as far as science believability goes.

They came up with Dark Matter.

If the matter was dark and hiding in undiscovered places within the Universe the dark matter could not be discovered because it was beyond detection. Then the flip side is if the matter was so dark it could not be discovered, then it was beyond being disputed. If you can't prove it is there, then you can' prove it is not there. This is foolproof. No one can disprove the matter being present because it is undetectable dark and so the matter can't be disproved because the matter is undetectable dark. Hell, this is the brainpower and the mindset of the smartest bunch alive, those that aim to keep the corruption of Newton ongoing and by conspiracy keep on fooling everyone alive.

But even the smartest minds alive can't fool all the people all the time. There is one small question eluding an answer. If the dark matter is there and has mass, then what is preventing the dark matter from pulling in the present. Why is the dark matter waiting for what to come to grip and avenge Newton by collecting the Universe and the lot to contract as Newton said it does?

The dark matter will somewhere in the future bring about a turn around and pull the mass for the Universe closer again, undoing the doing of the Big Bang, but if the mass is present and the mass does form a pulling force of gravity, why is it not pulling now and what is it waiting for before it starts the Pulling.

In astronomy and cosmology, **dark matter** is a hypothetical form of matter that is undetectable by its emitted electromagnetic radiation, but whose presence can be inferred from gravitational effects on visible matter. This is totally fiction and is as fabricated as modern science could be. According to present observations of structures larger than galaxies, as well as Big Bang cosmology, dark matter and dark energy could account for the vast majority of the mass in the observable Universe. This means if they can't see it they can't show it and that is brilliant to fool all the sceptics.

Dark matter was postulated by Fritz Zwicky in 1934, to partially account for evidence of "missing mass" in the Universe, including the rotational speeds of galaxies, gravitational lensing of background objects by galaxy clusters, and the temperature distribution of hot gas in galaxies and clusters of galaxies. Fritz

Zwicky is the "Father of Dark Matter," coining the term itself, as well as gravitational lensing and the sky survey technique.

He devised it but I can't say if he was part of the conspiracy or if his ideas were hijacked and misused by the conspirers. But in the end these ideas came in pretty handy to use in the rest of the conspiracy. Dark matter is believed to play a central role in structure formation and galaxy evolution, and has measurable effects on the anisotropy of the cosmic microwave background. All these lines of evidence suggest that galaxies, clusters of galaxies, and the Universe as a whole contain far more matter than that which interacts with electromagnetic radiation: the remainder is frequently called the "dark matter component," even though there is a small amount of baryonic dark matter. The largest part of dark matter, which does not interact with electromagnetic radiation, is not only "dark" but also, by definition, utterly transparent. Most impressive but here is the catch... If they are asked to show it they already admit they can't... because it is utterly transparent. If they are asked to prove it they already admit it is illusive and therefore they can't utterly transparent. They can make up the story on the trod as we run along because no one can prove them wrong because they can't prove they are correct.

The vast majority of the dark matter in the Universe is believed to be nonbaryonic, which means that it contains no atoms and that it does not interact with ordinary matter via electromagnetic forces. The nonbaryonic dark matter includes neutrinos, and possibly hypothetical entities such as axions, or supersymmetric particles. Unlike baryonic dark matter, nonbaryonic dark matter does not contribute to the formation of the elements in the early Universe ("big bang nucleosynthesis") and so its presence is revealed only via its gravitational attraction. In addition, if the particles of which it is composed are supersymmetric, they can undergo annihilation interactions with themselves resulting in observable by-products such as photons and neutrinos ("indirect detection").

This is the same as saying there are "anti matter" eating up matter. If they are asked to say what is "anti matter" or what is "anti matter" made of they can't say. It is a name and naming nonsense is a great Newtonian pastime. They do it to relax and to become social with other Newtonians, which may or may not be part of some mating ritual. If you named something you then it is as if you explained something because naming it and creating mathematical formula goes hand in hand. There is no need to be realistic because the end of any Newtonian intellectual capacity is to give a mind blowing mathematical formula of which the practicality remains a mystery and then give it a name. The name must be so impressive that just to remember it would take up all the effort any Newtonian has in reserve so getting to the point of proving it taxes the Newtonian's stamina beyond breaking limits and then there is no need for it.

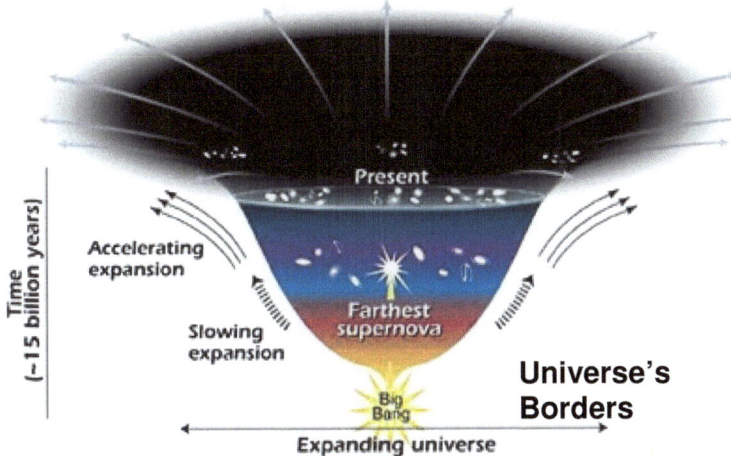

As the Universe developed from the Big Bang event, then the first "material" was apparently without "mass" because today it is hiding all the "missing mass" that will enable to Universe to conform to Newton's bogus claims of "mass pulling mass". See the black stuff that pushed out first that is the dark matter. What I would like to know is what is beyond the borders where the Universe apparently ends? What will we find in the space that has no space because the Universe ended just before the space that has no space began so ending the limits of the Universe? These question never are asked!

One thing it does not answer is if the **dark matter** does have mass it must have pulling power as gravity. Then what is the **dark matter** waiting for to unleash the gravity by mass to pull the Universe back into forming contraction. Why is it not pulling now if it is going to pull at all? Either it pulls by mass or it does not pull by mass but it can't have some retarding switch that will kick in at a time when it pleases the Newton's. What makes the **dark matter** slumber mysteriously while waiting to jump on the poor defenceless little Universe and force it to comply with Newton once more. No one can prove the **dark matter** is not there since no one can prove it is there. This is how one go about to devise a conspiracy. You keep it quiet and while everyone smells a dead rat but no one even thinks of looking in the right direction. The conspiracy is a success if everyone accuses anything but detect the true conspiracy.

Then science goes further and tries to detect the untraceable that is invisible. They try to find dark matter hiding in unseen places in a shady Universe. They spend billions on detecting non-existing dark material or dark energy research but not a single dime goes the way of finding out when the moon and the earth will have a gravitational self-destruction event. They tried to locate dark matter that was waiting to pounce on an unsuspecting Universe to start to pull it!

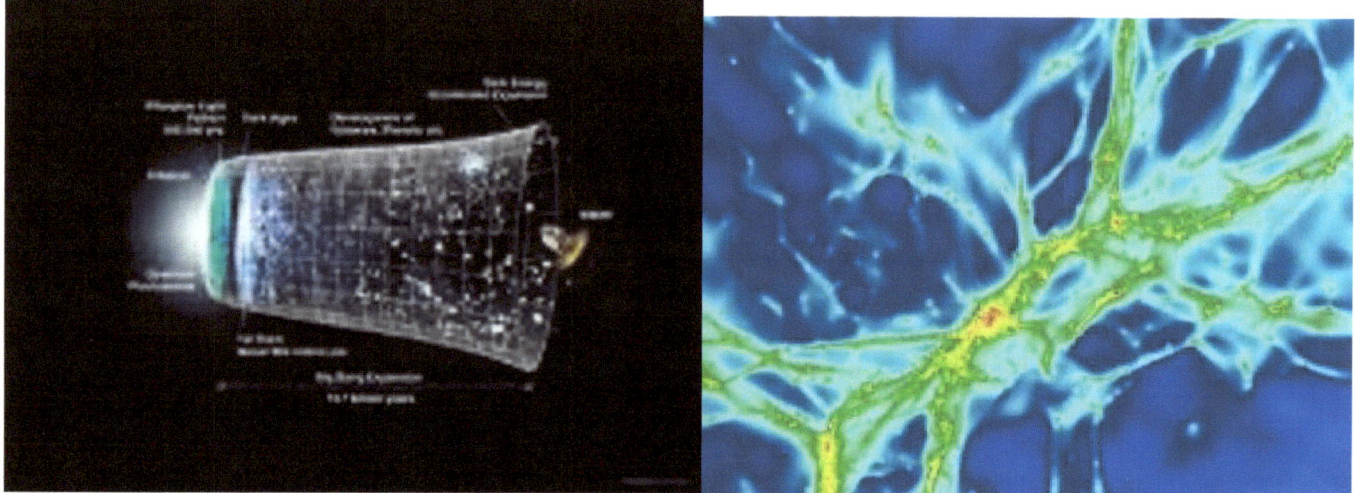

Do you not find that very odd…or is it just I wondering about what?

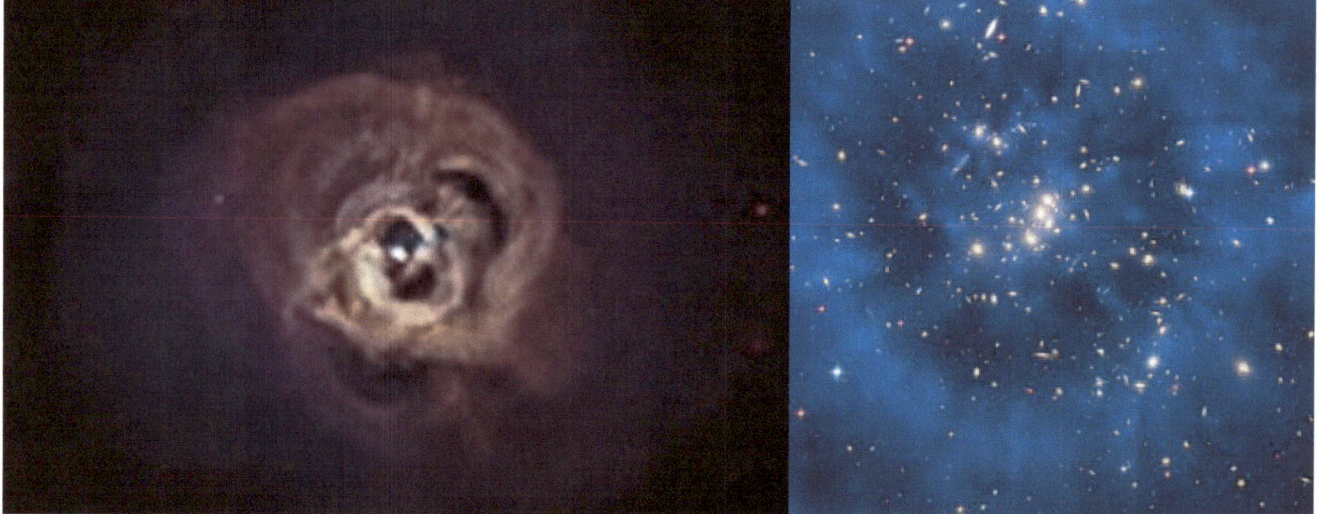

All the above pictures now are evidence of dark matter. Can you see any dark matter? The question science has to answer is if the matter is dark why has it not got any mass because according to them this is the part not of missing light particles but missing mass. It is where they hope to find the absent mass to make Newton's hoax work. If it is in place why does the mass not pull as all mass supposedly should do?

Someone in science thought it was worthwhile to study what would happen in the very solar end when the sun comes to a final exploding conclusion. Someone in science took time to bother with the end of the solar system and what will be the applying conditions during the finale era of our sun having a solar system. All these elaborate mathematical conclusions is merely science fiction promoted to cover-up because the sun applies gas and it is not burning coal in a furnace or boiling water in a steam engine.

They even measured how big and how red the sun would be in the end but never do they bother about what is more obvious, if Newton's formula is correct then they should calculate when will the earth and moon get together. That seems very odd and suspicious to the less informed…is it because they knew already mass wasn't pulling anything in particular to anywhere? Think of what effort it took and still takes to find the reason why the earth is growing in size as it does and why the space between the earth and the moon is expanding by having the moon growing more apart from the earth? …And what human effort will go into understanding all the reasons why that is. They are relentless in their quest to find answers but only when it is to cover misconduct or to instate the correctness of what can never be correct. Yet, ask them when it will come to the time when the moon and the earth will finally collide because of Newton's claims on mass pulling to bring contraction and we find only silence. It is the sort of silence I describe when I point out what to look for when one goes in search of a conspiracy. I find not even a whisper…in astronomy and cosmology, dark matter is a hypothetical form of matter that is undetectable by its emitted electromagnetic radiation, but whose presence can be inferred from gravitational effects on visible matter. This is totally fiction and is as fabricated as modern science could be. According to present observations of structures larger than galaxies, as well as Big Bang cosmology, dark matter and dark energy could account for the vast majority of the mass in the observable Universe. This means if they can't see it they can't show it or disprove it and that is brilliant to fool all the sceptics.

One thing it does not answer is if the dark matter does have mass it must have gravity as pulling power. If it is and it has mass it must pull this very instant and that will bring contraction because that is what mass does. Then what is the dark matter waiting for to unleash the gravity by mass to pull the Universe back into forming contraction. Why is it not pulling now if it is going to pull at all? Either it pulls by mass or it does not pull by mass but it can't have some retarding switch that will kick in at a time when it pleases the Newton's. What makes the dark matter slumber mysteriously while waiting to jump on the poor defenceless little Universe and force it to comply with Newton once more. No one can prove the dark matter is not there since no one can prove it is there. This is how one go about to devise a conspiracy. You keep it quiet and while everyone smells a dead rat, no one even thinks of looking in the right direction. The conspiracy is a success if everyone accuses anything but detect the true conspiracy.

When we go in search of what principles applies to form the building material in the Universe we better look and see what is it that the Universe shows us most graphic and we better stop telling the Universe what it is that we want to see and what the Universe should offer us that we wish to see. We better stop telling the Universe it must get mass and start to see what the Universe tells us what it has to offer us to see. If stars burst by releasing heat then stars are constructions that confine heat or cram heat into a small space. If this is true then gravity must be the process of freezing heat by turning movement and displacing space into compacted heat making gravity a process whereby space freezes as it condenses.

Looking at the Universe in singularity from this angle as Newtonian science portray the singularity picture the idea seems most beautiful and by applying the magic of mathematics it seems to be so real except for one problem we encounter and that comes when we use a dash of logic to ask Newtonian science for some straight answers. When using equated mathematical formulas one include a certain part of the Universe by excluding the rest of the Universe.

There is no formula that has the ability to contain the Universe in its inclusive entirety because the Universe is eternal no matter how one would look to appreciate what one sees. Looking at any part of the Universe the distance we don't see removes the quantification of the picture that we see. What ever forms a picture of the Universe excludes untold many, many times more than what the picture we see reveals because it is shear stupidity to think what we see in any picture represents everything there can be.

This picture present the thinking of the most supremely intellectuals God ever put on earth. Their ability to formulate and philosophy about cosmology in their eyes and in the image they portray equal that of God Almighty. That is why they are all atheists. It is because they can't believe there could be anything more supreme intellectually that what they are.

Sit back and see how this illiterate boorish labourer and complete novice which is what they all think of me and what they like to call me show you the understanding ability and the brilliant reasoning ability of the atheist. See how pathogenic their intellect is when they put their abilities on par with God!

See the waves. They see the waves as forming singularity. Singularity is what it says; it is single dimensional. If there are waves it is the first sign of being multi dimensional space because there has to be space in which a wave can roll. The wave has to be higher on certain points and much lower on other points to form a wave pattern. This wave then calls for dimensions and singularity has no dimension but forming a singular concept that excludes space. The mere fact that they include waves shows their ability to fathom singularity. I repeat; this is the absolute nonsense you get when using mathematical equations from which one derives cosmic ideas and to think that lot feel smart arriving at such conclusions. The reason why I hammer so much on the mathematical instability equations deliver is that in the book/s I show that my articles are not published about the Absolute Relevancy of Singularity because I don't use multiphase mathematical formulations. This is what is said and I quote In my article to Annalen der physics I used 15 pages to explain this process of singularity applying. I received a rather cordial but sincere reply from the Editor of the magazine. When I placed an article in Annalen der physics Dear Prof Friedrich W. Hehl said in the e-mail he sent me that there is no way to "explain" the world of physics I am not going to go into detail how this works.

On the other That side of the Pythagoras's' triangle we have 1 going square. That makes Pythagoras's' triangle 49 + 1 = 50 on the one side of the earth and the same on the other side of the earth. The total is 100 and the square is 10. That leaves the Titius Bode law with a value 7 (it forms part of the material of one body) and 10 in relation to the space. Then from the relation of 7/10 and 10 / 7 forming Π the Titius Bode law form Π^2 applying "With a lot of words and some simple algebraic relations" to quote Friedrich W. Hehl, Inst. Theor. Physics of Annalen Der Physics fame. This was simple algebraic relations but still it is science, is it not?

This is the notion Einstein and his group of wise men had about his mathematical formulating of singularity. Let's see how they see what singularity is and don't get me wrong for it is not singularity I attack but it is how they and their mathematics come to conclude singularity. Singularity is a mathematical dimension of 1 being 1^0 or 1^1 or 4578^0 or Π^0 but the final value can't exceed the total value of 1. Having any number above 1 goes into space and singularity represents space less ness. If we look at this grid that Newtonians form their opinion about what is the position on singularity we see a form based on what looks like some Scottish kilt of some sorts and the kilt we see has waves.

If there are waves there is a depth and if there is a depth there has to be space within this grand picture of singularity without space. Space has to be three-dimensional because space has to have six sides. If I can see the topside there has to be a bottom side because the topside begins and the bottom side ends whatever fills in between top and bottom. With a top and a bottom the sides must end to have the bottom separate from the top and then it must have a left and a right-hand side.

That proves they put substance in between whatever holds the topside away from the bottom side. That then is material and not singularity. If the grid criss-crosses then there is space because to validate criss-crossing grid pattern requires space and in singularity "where the Universe goes (flat?)" there can be no space so how can they with all the brains between those oh so wise find a dimensional grid with waves and all. Then when I explain singularity by involving mathematical principle I am scowled at in the phrase.

The Newtonian mathematical stupidity starts with having the idea that a line or a graph starts with zero. If it starts with zero it never can end because it can never accumulate that with which it started. Whatever a line starts with the same value has to repeat in order to get any mathematical sequence.

Start a line with a dot to the value of zero and this happens: 0 + 0•= 0 + 0•= 0 + 0•= 0 + 0•= 0 + 0•= 0 + 0•= 0 + 0•= 0 + 0•= 0 + 0• =0 and then start the line with singularity continuing to form a Universe this is the result. I would love to see what is more basic in mathematics than the mathematical line by compiling the line as follows:

1^0• + 1^0• = 2 + 1^0• = 3 + 1^0• = 4 + 1^0• = 5 + 1^0• = 6 + 1^0• = 7 + 1^0• = 8 + 1^0• = 9 + 1^0• = 10••••••••••

Mathematically singularity is anything and everything to the power of zero excluding the number zero and please informs the wise about this fact. On this principle I found the way to prove how the Universe started before all material that was already present started the Big Bang process. This is mathematical reality. Whatever the reader brings to mind must not be of cultural bias because whatever mathematics brought to mind and whatever physics brings to mind the dimensional quantifying does not exist in the realms of singularity. That which connect the Universe are innumerable lines all connecting by criss-crossing as circles representing singularity. This is what made me realise how the Universe started.

There was one • that became one dot • that that duplicated in size •1•2 and the movement of duplicating what is singularity became two •1•2 dots that became three •1•2•3 dots that became four •1•2•3•4 dots that connected as five •1•2•3•4•5 dots after which six •1•2•3•4•5•6 dots came about and then seven •1•2•3•4•5•6•7 dots formed a line••••••••…but it was a line and the line had dimensional references as it still has to day without having dimensional qualities as it has today by forming space.

This is the essence we find when finding a triangle has the same value as a straight line and the half circle has equal value to the two factors mentioned. The line always refers to two•1•2 (the half circle coming from movement) showing relevancies not only from where it came which is the triangle •1•2•3 but also putting references applying to where the line is going because while being two the being part represents the triangle that represents the three of two going to two 2^2 bringing about the movement taking 1 going to 1 (1+1 =2) but also going in the opposing direction of going forward. This produced the double half circle •1•2•3•4. This is the basis by which the Universe started and remains movement. These pictures show movement!! That proves no line can start with zero and no line can grow by zero.

It is one thing to be brainwashed and another to allow it. Physics students, it is your duty to pull the plug on the powers of the All-Powerful Academics in Physics and stop their dishonesty. It is your task as the as the next physics generation to stop the criminals that are filling the corridors and the lecture halls of physics departments throughout the world by acting as if they know all there is to know and all they know is to fool the next generation of students. Stop their teachings by forcing them to stop their criminal fraud.

Force them to explain the deception such as the one they call THE CRITICAL DENSITY, which is a conspiracy to commit fraud. Let them explain how an expanding Universe can suddenly and abruptly turn in direction of developing and start to contract as Newton stated it is doing at present, and when facing all other concluding evidence showing that the Universe was expanding since time began they come up with the utmost unrealistic garbage only an idiot can devise. Tell them to bring proof with evidence that the cosmos is contracting as Newton said because from Hubble's evidence physics became a hoax.

In THE CRITICAL DENSITY conspiracy all they say is that they are waiting to see when the cosmos would stop its criminally insane behaviour and start to listen to the laws of Sir Isaac Newton. They shove all the blame of wrongdoing onto the cosmos and take away all error from Newton. If the cosmos does not contract as Newton said then when will the cosmos mend its ways and follow what Newton said and to start contracting! It is a conspiracy to cheat and lie and crook the human race in order to keep Newton untouched. If they get rotten egg off Newton's face physicists would not have to share Newton's stink.

With The Critical Density shambles the modern Newtonian set out to defraud the world in the same manner as their **Master Sir Isaac Newton** has done centuries ago. Newton said the cosmos is contracting. When Hubble proved the cosmos is not contracting, Newtonians looked where the cosmos went wrong by not following Newton guidelines he so clearly set the cosmos to follow. It has to contract and not expand. Those in academic positions fabricate non-existing material no one can detect to cover the real conspiracy they try to hade. When the argument arrives of contraction versus expanding they wall this down by referring to the search for a substance that can't exist and could never be detected. It is not the dark matter issue that is the real conspiracy but the dark matter forms a conspiracy to hide the facts that the true conspiracy covers up. It is this mother conspiracy I am gong to uncover and present.

I realised I have to reveal what they hide in order to promote my findings and the only way is to come clean but doing that is harder than it sounds initially. These guys I take on has clout, money, power and influence. This letter is about finding investor/s in my attempt to publish a book that no one in science wants to have published. I am revealing what is hidden for hundreds of years and I challenge any one to prove what I reveal is not a conspiracy or that I am wrong about science being wrong and that they hide this incorrectness as hard as they ever possibly could. What you are about to read is the mother of all the conspiracies in science. You are going to read about how science applies a system of mind control with thought processing and while you are reading this remark you are also a victim of this brainwashing process we all endure.

All conspiracies Science ever produced such as the Critical Density formula or the Dark Energy Concept, or any conspiracy connected to science is in place to protect this fundamental conspiracy. Everyone outside physics must know the truth about my effort to reveal the conspiracy that those teaching science hides. All other conspiracies hides the Mother Conspiracy so effective that in three centuries no one outside science got a sniff about what the Mother Conspiracy entails.

The Mother Conspiracy is in place so that students in physics are brainwashed by instigating the deliberate sanctioning of mind control on students through their practising of enforcing thought control they unleash on students. Sounds unbelievably harsh but when you completed my book you will realise how they do it. Sounds like another nut-case trying to claim one second of eternal fame on the account of much worthy gentlemen that only fights to promote honesty and truth...yea well read this what I have to say first and then draw your conclusion as you see fit!

I prove that the Mother Conspiracy is in place. I reveal the Mother Conspiracy and I show why it is in place. I show how every student including you reading this has been brainwashed to believe science is believable. Download this letter and get the nastiest surprise of your life. See how much you are brainwashed. Then also read how the cosmos truly applies science and read explanations of what was never explained before and how the explained factors interlink. The brainwashing is continuously repeating what is impossible and so many times it becomes believing it as trustworthiness every person on earth shares. They join some everyday facts as truth with a total lie and hold the truth out for all to see while clouding your perception to make you believe. Then they hide behind a lie that is out of sight and out of mind but forms the basis of everything we all believe.

I wrote many books that went unpublished but I wrote seven articles, which I call the Absolute Relevancy of Singularity in which I go much further than Einstein did on the relevancy of singularity in the general application of all types of physics. In doing that I had to reject Newton's claims about gravity. I prove the Universe is made up of singularity and contains only singularity in many forms thereof. I prove singularity is not a general phenomenon but is the concept forming the absolute basis of physics and this foundation is transferred to the way that mathematics began. The Universe is singularity and proving this is simple. It is the other concepts flowing from this that complicates things.

Modern Science can't explain basic things because of a lack of proficiency. After finding the building blocks of the Universe I now can show where is the centre of the Universe? Have you thought where the centre of the Universe is? I can answer that...and why is the Universe still growing since the Big Bang...why did the Universe start so small...why did the Universe fit into a neutron at one time...how did everything expand from fitting into a neutron...why does space grow from small to large...where is it going while it is growing...why was the Universe any specific size...what was everything before the Big Bang...if you read this book I propose you will know all of this!

My discovery is the fact that singularity presents the total and complete control of everything there is in all forms of everything there ever could form in the Universe and discovering this fact led me to prove that gravity is the application of Π and from extending Π further into a six dimensional sphere that is spinning in a six-sided cube, is what makes the Universe form three dimensions using time thereby the cosmos enlists the multi dimensional sphere we live in. Everything in the Universe is round and roundness is Π.

However, at the stage where the Universe holds gravity as Π, which is at the point before it forms multi-dimensional space, the Universe still is flat although the Universe then is already accepting form by going round in shape-forming when extending from forming singularity as a value when retaining form in the principle of a sphere. This at this stage might sound bewildering but with better explaining it is rudimental. I only have to show where to find singularity to show how simple it is to understand. I do not wish to burden you with more information about my work at this time because I require your attention on another matter first. I need a backer to fund my work but first you must read on and see how unique my work is before deciding. My findings are not congenial with centuries long accepted views so no one in Mainstream science could dare to look at my work. If they do they have to confess that Newtonian science is based on outdated views that hide most important issues and is a fantasy, it is not the truth.

This past thirteen years plus saw me go without any income as I tried to get my theorem recognised as well as get my warning noted. Going without a steady income left me almost destitute and in order to find a manner to get my theory across to the attention of influential readers, I decided to publish my theses of eight books electronically as to try and get around the stranglehold of Newtonian bias controlling science at present worldwide. I decided to publish electronically which those in power do not control. However the market out there is a jungle filled with others also trying the avenue I chose making the lot of us dumped in a swamp of books, all of them vying for customers just the same as I do. I decided I must make a plan! What you read as a letter looking for a promoter is the plan I devised. It is not bright but it is all I have left.

Science forever hides the truth of any subject behind this veneer that the public is too stupid to understand and therefore they disclose their opinions about whatever field they wish to promote with no explaining how they came to the conclusion they have. Not once ever have they lifted the cover of what they hide and about all the incorrectness of what they say. Every person on earth takes what science says in declarations as truth but because it is culture to accept their word as the final truth on all matters they never admit or divulge any incorrectness on their part.

It is the information about the unbelievable oversight of Newtonian mistakes I disclose how they keep the oversight of Newtonian mistakes silent and why they don't divulge that which they keep silent about. It is about them never committing to the entire story saying what all sides of the subject may offer by giving an all-round presentation of everything anyone would require to know as then to be in a position to evaluate with insight. This idea inspired my effort to find a partner that will fund the promotion of there books or a single book started off as I realised it was about time to inform the public about a science conspiracy.

I can't get published because I don't acknowledge the correctness of mainstream physics and I say it out loud. I found the most important information anyone ever discovered and science doesn't care because what I found trashes what those in science believe is true and condoles as the truth. They frustrate me to the rafters with their blocking of my work getting published directly or indirectly. I must inform the public about the conspiracy they ardently cover. At first I started by supplying little information but soon realised to get my message across I have to supply more information about what forms a comprehensive study on cosmology. You may observe while reading this book that it seems as if my frustration will ring through like the chiming of the Big Ben Bell. For that there is a reason. At times my frustration and anger will boil over drowning my politeness and that is true, which I admit. For twelve years I have had the answer that would correct the philosophy that has a stranglehold on cosmological science. I discovered the building blocks of nature where my discovery puts all other cosmic aspects of science into science fiction. But because my findings don't salute Newton's ideas, my finding the Cosmic Pillars are buried.

I have done a study that took me since 1977 on physics and in 2000 I had enough information to present a complete theory on explaining the cosmos in a much more sensible manner. With al the facts I concluded in my decade's long study I started to look for a publisher. To my shock I realised that to find a publisher willing to take on mainstream physics is not that easy and to get published.

I began by trying to make Newton work but I couldn't. I went to the professor so many times but he only explained the formula but never put the formulas to work. Then when I couldn't make sense of these formulas he put the blame on my mathematical skills hinting that I might be too poorly equipped to handle Newton because after all it was Newton we were dealing with and Newton was the most brilliant physicist that ever lived. It was rather blatantly pretentious of me trying to match the skills of a man being in the class of Newton. However at that point I already had my doubts about Newton and was very little impressed with his statement of $\frac{dJ}{dt} = 0$ which is a mathematical impossibility. This is simple to prove. If you divide anything by zero it is zero. If you divide 2 / 0 it is equal to = 0 or if you divide 3 / 0 it is equal to = 0 or 5429856 / 0 = 0. Then on the other hand if you multiply anything by zero it is just as much zero. If you multiply 2 x 0 it is equal to = 0 or 3 x 0 it is equal to =0 or 5429856 x 0 = 0.

I don't care how much a genius the man was but if he said anything divided by zero had a value he knew very little about mathematics and I was never that much impressed with his skills. Newton said because of the value of $\frac{dJ}{dt} = 0$ which is a relevancy he showed that apply when the top spins to get the top from where the top was motionless to where the top is spinning. If it was zero then no top is spinning because

$$\frac{dJ}{0} = 0 \times dt = 0 \quad \text{or} \quad \frac{0}{dt} = 0 \div dJ = 0$$ and this is the mathematical truth about mathematical

laws applying. Even Newton can't change that and it shows how little did Newton know about applying mathematics

$$\frac{d}{dt}\left(\frac{1}{2}r^2\dot{\theta}\right) = 0, \qquad \frac{P^2_{\text{planet}}}{a^3_{\text{planet}}} = \frac{P^2_{\text{earth}}}{a^3_{\text{earth}}}.$$

Keep in mind when Newton concluded that and also no one ever thought these values would be discovered. Newton was on very safe ground because he knew no one would ever detect his misjudging of cosmic principles. The Newton formulas were impressive and scary to all that did not know better. Fortunately at the time the information was not available to prove the formulas correct.

$$\left(\frac{P}{2\pi}\right)^2 = \left(\frac{a^2\sqrt{1-\varepsilon^2}}{\ell}\right)^2 = \frac{a^4(1-\varepsilon^2)}{\ell^2} = \frac{a^4(1-\varepsilon^2)}{a(1-\varepsilon^2)GM} = \frac{a^3}{GM}$$

Try to implement without knowing how far the planets were, how many planets were there and most of all how massive they were.

Please take note of a conscientious warning about the gravity of the misgiving there is on the part of the most respected Academics in physics about a much concerning matter.

I state it emphatically that science accuses me to be not schooled to the point where I am able to have any form of an opinion on any matter concerning Sir Isaac Newton. Notwithstanding that my research proves I did my private studies and through which I skipped the indoctrination and mind control academics place on students, goes unrecognised by their standards and so too my ability to have any insight on matters regarding physics.

I challenge them on the other hand to make Newton's concepts about gravity work. It is one thing to know everything there is to know but if everything you know is about fiction and not facts you might just as well know nothing because then you know everything about nothing and nothing about the truth. I have brought science again to a crossroad as it was days of Kepler and Galileo where they both clash head on with Newton and never did Newton in any way compliment their ideas. Newton says mass pulls, which must make objects fall unequal because of different mass. Galileo said all things fall equal, not a feather and a hammer in a vacuum but everything we see on TV falls equal notwithstanding mass differences. That strikes Newton's ideas of mass pulling mass out the ballpark.

However, my skipping their methodical and systematic brainwashing enabled me to see and allowed me to be able to express the incorrectness in Newton's teachings and allowed me to show in clarity what destructive force Sir Isaac Newton used to corrupt the laws of mathematics, corrupting science along the way and mostly raping to the work of a great man, Johannes Kepler and what Sir Isaac Newton did can only be expressed as being blatant criminal fraud. What his deeds amount to is to corrupt the laws of mathematics, to render the laws of cosmology useless and to rubbish all of science. Should you find this to be unbelievable, then I am glad to announce that this book is more for you than any other person, so go on and read what academics guarding science never wanted published.

I challenge anyone who disputes any claim I make to prove me wrong by proving me wrong and not merely suggesting claims in that direction or to declare Newton is proven so Newton needs no proving. I furthermore challenge any person to prove the solar system or indeed stars form in the manner of mass as the Newtonians teaches. I challenge every person to prove that the Titius Bode law is not the way the solar system forms. Here it is in a nutshell; what this is it is not a theory that is called the critical density theory but it is a criminal venture conceived by the intellectual minds in physics to go criminal and defraud the world. They decided that the blame of the Universe not working as Newton said it does has to be placed at the door of the Universe. The Universe is lacking matter to contract as Newton said it should. Newton still remains absolutely correct. The blame for the mistake is diverted to the cosmos and away from Newton. Newton did not lack the insight to see the Universe expands, no, the Universe lacks material to contract. The Universe made the flaw and now we must find how the Universe will correct its flaw. The Universe went wrong by not having the sufficient mass and therefore the Universe must correct the lack of mass to get the Universe back in line with Newton. The blame for the mistake must be laid in the midst of the material within the Universe. Newton's formula stands correct and all suspicion goes in the way of how the cosmos was designed not to applaud the truthfulness of Newton.

Whatever they dispute or however they ignore me I am the only one since time began to explain as much as prove how the solar system forms as I explain and prove how the Titius Bode forms. But the Titius Bode can only form by implicating the Roche limit, the Coanda effect and the Lagrangian points and notwithstanding the importance of my uncovering of science principles that were never yet understood before science and Scientists ignore me and give me the cold shoulder. I can explain gravity for the first time ever and in science no one shows interest because if they give me any credit for being correct then they have to admit that all the information they approved so sincerely was falsified facts implemented since the time of Newton and think of all the rotten egg they then will have dripping from their faces.

I wish to explain a small part for what my work entails and that is not in the form of information that I present. There was a part of my life that I slept one hour per night and worked for twenty-one hours per day to find answers. I have this ability to concentrate beyond the normal when all my senses of functioning in a normal capacity is drowned and all that I then can do is concentrate on coming to a solution. To formulate the science that I concluded I slept for one to one-and – a half hours made time for eating and bathing for one hour all together and spoke to my children for about one hour just to keep connected to reality. This information is not even remotely on the same level as what I present in this offer because what I formulated is very complicated and requires a very highly developed level of concentration just to follow the mathematical arguments required to read the work. It is a process of calculating relevancies about how stars develop and what goes on in every level of growth through which a star develops. That is one line and there are numerous lines of calculating the cosmos.

This twenty-one hour days of work went on until I got an eighteen hour long heart attack (I never even knew any heart attack could last that long) which led to a nine hour open heart surgery after which I sit with a heart less than half the size it was before. That brought about a nine-hour open-heart surgery where I was "frozen" for seven hours. Any person who had some heart condition will be able to realise the severity of this long theatre open-heart surgery operation.

I say this to indicate it was much more than money I sacrificed to gain the knowledge I present in my Theses. What I got in my self-study opportunity did not come cheap. I almost lost my wife and my family in the process but fortunately before I lost my wife and family through a divorce I lost my health and that put a stop to my excessive working. I lost my farm and all my property in the process of battling for answers in science but moreover I lost my health and that is a much bigger loss. I admit I was beyond obsessive but that is the only way to be to get the conclusions and the corrections in science that I got.

It is that which I try to share but I am frustrated by Academics who know their approach to science is completely flawed and that they hid it for centuries. I got the answers in the end but wrote off my health. I am almost in desperate health because I gave too much when I fought for answers. Today I am without pension or any income because I just can't work being a diabetic with a heart condition suffering from depressions but this last part is because of our financial position. My wife is a typist in a government school in Africa and she gets about (in British pounds) £204 or bout $306 per month. In any country that money is not enough to go around in comfort.

I mention this to say that what insight into science I may have received through my work we as a family endured much suffering to allow me the privilege gained by that knowledge. Not one bit came cheap and the price is just too high to allow a bunch of cheating academics in physics to undermine my work because they think they are God on earth they judge my work as being heretic. Today if I want a cold drink I have to ask my wife because with as little as it may be she is the only one with an income. That is why if I can get one book to sell in bring in some funding to help boost our income with any amount it would make our everyday position more bearable. This then gave me the idea to write commercially motivated books and get a **Print-On-Demand** publisher to bolster my income in the books because at present there is no income in any book. There will be three hands that collect money. It will be mine, the publisher that fairly takes its cut and the retailer doing the funding of the promotion of the work.

I have a choice between tearing up everything I worked for during my adult life or then to be more precise ever since 1977, which was the time when I realised I can at least try to get some understanding about what in science is missing. Then following that choice one lifetime of labour and all my fighting for answers while going through nights and days without sleep will be gone and the misery of hiding the truth will then win and fighting for the truth would be lost. That rather don't entirely fit my persona and that is not me and so I have tried (much in vane) to side step that part by publishing books where at present these books drown and are flooded by tens of thousands of other books also vying for a market.

My second option is that I can sit back and let God's grass grow over God's acre and get grass to cover my work that I published on the Internet and hope someone in eighty years time will come across it, understand it and see its worth. Then that person would come as Newton did with Kepler's work, not have a clue what the hell I was talking about exactly like Newton did with Kepler's work and rape it, bash it, try to kill it and make a jolly mess of it just like Newton did with Kepler's work because I will be dead for eighty years or so and that person will be able to read it just as Newton was able to read Kepler's work and make no sense of what he reads exactly the same as Newton did with Kepler's work and make such a mess it would become a joke just as Newtonian science see Kepler's work as a joke. Allowing this option to come in place will allow me to let those ego driven narcissists cheat the public by their falsified lies for many years to come. You might be shocked by such harsh words but even at the end of reading this letter you will see a very small part of all the facts those physicist lie about because they feel their position is above approach and beyond reproach. Thus if I leave it to be then the tendency of Academics prevailing at that time will favour my work because the trend would have changed and science will then be such a mess it would float in its own disposable waste as it does today. The difference in today and eighty years from now would be that science would choke in the riddles it now hides.

My third option is to get some form of income by allowing someone or some persons to share in the income of my work. If I could sell some small part of my work that is very little informing and even less complicated I might be starting a movement to get people thinking and asking the correct questions about what is clearly inadequate backward thinking.

What do you think would the reasons be that I am so explicit in my deliberation about the Critical density idea and detailing the fabrication of dark matter and lost mass and all of that? If they went that far to protect Newton by devising billions of dollars promoting senseless ideas just to maintain the image of holding Newton up as the most believable person with the most correct visionary ideas that ever had human intellect, then think what they will do to me and my work to protect Newton where my vision completely destroys Newton and where I prove why the cosmos does not apply Newton but does apply the four cosmic pillars or the four cosmic phenomena that forms gravity and without these four phenomena there is no cosmos. Science at present will never accept my views because my views don't include Newtonian science's fabrications and I have to deliberately exclude all of the Newtonian religiosity and mythology.

It is essentially very simple ideas from which a bigger picture grows and it consists of questions, which no one ever thought to ask. By never asking for proof about Newton those powers in physics cheating the human race with falsified facts are getting away with intellectual murder. I include four options of books, which are prepared for electronic publishing. There is one that is prepared for printing purposes.

The books all carry the same topic but the information is considerably less or more comprehensive. I include a selection ranging from everyday matter of fact to rather academically motivated and although the information is easily digestible to somewhat informing it caters for a wide range of intelligence.

In every case each one asks for a higher level of concentration or a much higher in depth understanding level than the one before or the one following because to inform and to give backing by proof requires a lot less or more concentration of factual information. The facts presented is the same but the degree of being presented with comprehensive information does alter the purchase field that the book will enjoy. However, for the sake of explaining the contract I have in mind I will stick to the use of one title but all titles will be offered in a deal on equal terms. It is the prerogative of the investor to choose which market would the investor feel he would receive the highest return on an investment.

Should I be so fortunate as to find more than one investor then I could offer any one book to many investors as to offer a range as wide as possible in order to promote my new cosmic theory. I sit with a predicament that I have to write in such a way as to allow every person an insight into how I present the cosmos and all the while I am in a fight with the establishment about what I show and that I have to prove what I say constantly. It is always a balance in what I say and how I prove what I say without scaring the daylights out of the readers or being blown off the table by academics. Nevertheless let's use the name **A Science Conspiracy to Conceal a Cover-Up** to use any name as a name for the books on offer.

People who are in a mindset of getting information for free normally have not a large capacity to analyse information on a higher level. The more a person is prepared to pay for information the better quality information the person demands because the more will such person mentally be developed. A person who downloads information available for free will be satisfied with information that is on par with the details equal to what is available on TV news or other very ordinary information.

To expect of such a person to be able to understand information that is of prime quality is asking for far too much. Such persons form the gross majority of society and where they feel a need for information, what they ask for and the level they can observe is normally that which goes for free. When a person only pays $6 .00 for information such a person realises that is what he or she can get and normally can divulge. These are plentiful but when we go into persons who are prepared to pay $200.00 for theses in physics they are the ones that can absorb lots of information but also they form the very small minority of say less than one present at the very top. There is a market niche and there is a limit of expectation to meet the requirements in delivering the expectations. This puts certain readers in certain market brackets that one has to cater for.

In the field of your expertise namely publishing or then helping me finding a sponsor or investor that would enable me to get published I hope that you know of any investor that is not scared of taking a chance and help me to take on the academics ruling the world of physics. I am sure you would know of someone in the field willing to engage in a fight standing next to me. I whish to find an investor that will promote the e-book at that partner's expense and share in the income the book generates.

In the printable option such a deal can also be struck where the book is published by Print-On-Demand and I get fifty percent of the profit and the person paying for the print and doing the promotion and distribution gets the other fifty percent of the profit. The costs can be deducted and the rest of the profits shared evenly. However this will not include the academic books that have very little scope in the commercial market and all information remains not for sale or not included in any partnership. It is an open venture and I am open for suggestions as long as it is fair to all sides. However I retain my rights on the subject without compromise or sharing my information on any grounds. The theorem and all such information is mine unconditionally and remains mine and can never be sold or changed in ownership.

My books are special because I show information about science that no one even ever suspected let alone are able to prove. I show that modern science is a hoax and a folly and I dare any person layman or academic master to prove otherwise or to prove me wrong in even the slightest detail that I present. I am turning to you because I need one or more investor to help me publish my book since you will see that I am the only person who has the guts to confront those filling very important places and those covering their fraud with very important offices that protect their positions that give them undeserved dignity. If you think this mud slinging is big words and much boasting then I inform you that compared to the truth I uncover it spells out controversy as never before. Think how big will any controversy be when someone proves Newton's idea about mass pulling mass is one big hoax that was and can never be proved and the entirety resting on such a claim is bogus science.

Think how big will any controversy be when someone proves Newton's idea about mass pulling mass is one big hoax that was and can never be proved and the entirety resting on such a claim is bogus science. Controversy and scandal makes things sell but in my case those in charge of astrophysics smother me and I am silenced because they are so powerful they could silence me up to now. To be published they must underwrite my work and that they are not prepared to do because I bring information that puts everything they say in dispute as much as disrepute. I am surprised about what flurry the moon landing controversy holds because what I have to show overshadows the moon landing controversy by light years. Considering what hoax I present makes the fact that the moon-landing event ever took place or did not take place much insignificant and quite pale in comparison.

Should the moon landing be a hoax, then one department in the U.S.A will be damned by the public for one presentation but the hoax I uncover puts the entirety of physics concerning astrophysics in the open as the biggest hoax invented by man in the history of mankind. The fraud (and I call it fraud for there is no better description) that science hides from the public overshadows the controversy about the moon landing by many miles. I have worked on this subject that I present ever since 1977 and I also bring the solution to remedy the matter. Yet no one in science wants the public at large to read my work and be informed about the many details that I uncover and what I bring to the table to solve the unproven dogmas they present as fact because then they have to admit there is a problem with their science and that it is based on fiction and not on facts.

These are the facts I found applying in the cosmos and with these phenomena working in nature and Newton telling the lot out there are forces that works on "mass" pulling "mass" it makes science fiction. If something is not applying and I tell you it is applying I am not conveying science but fiction. If I tell you cats fly away from birds trying to run the flying cats in it is fiction. Although Newton has never been supported by cosmic evidence, still everyone is sharing the Newtonian vision of a contracting Universe where the lot would one day come together and Creation will end where Creation started some time ago.

The supposition is that Universe has ends and the ends are drawing closer by the mass that is pulling mass towards one another and we are in the centre of an ever-shrinking Universe where this process is named the Big Crunch. The earth pulls us closer while the earth pulls the moon closer while the sun pulls the moon, the earth and us closer. That is what the lot of us can see… we are forming the centre of the ever contracting cosmos where every Newtonian can vividly see with his or her eyes through any telescope that all Newtonians minded scientists are sharing the centre stage of the ever collapsing Universe. That is the conspiracy holding science at ransom and caged in, locked in a cocoon of ignorance for almost going on to be three centuries.

Try as I may while no one I approached can prove a force called mass has pulling power, I could convince no one there is no mass anywhere because there is no such evidence, but in that is the devastation of the conspiracy. I am fighting a religiosity called science based not on fact but on accepting culture with a head priest called Newton and a cult or sect that has a demonic hold on the minds of the masses. If you believe in mass (not how much you weigh but mass that pulls) then prove it.

All I ask any one in science is to show where does nature within the cosmos apply mass which then forms gravity to pull anything towards anything while expanding. They put mass down as the over riding fact and yet nowhere in the Universe is mass upheld by the Universe as an accepted phenomenon. Forget about the fanciful corrupt mathematics that proves nothing when the cosmos does not confirm Newton's crooked mathematical arguments. Newton's religiosity might corrupt science but who in science would cares about correctness when it simplifies the ongoing brainwashing of students studying science. They confuse everyone about what *weight* is, what *mass* is and what *gravity* is because they wish to have everyone think of "mass" in terms of weight while they then deny weight and "mass" is the very same thing and then they confuse "mass" and gravity because they never distinguish between what "mass" does and what gravity does. This is only the tip of the iceberg and you will see when reading this book.

This is what there is and that is all there is. The measure of mass forming gravity clearly plays no role in allocating the positions of planets as Newton declared it must do. The entire idea that gravity is a magical force created by the value of mass is as unbelievable as the dogma is of those presenting this idea. Please use what the solar system provides to confirm what Newton says is in place when he says mass forms gravity. Science would rather accept Newton where there is no proof of Newton ever being correct than to admit Newton's incorrectness. Science would rather deny there is cosmic principles that is in place in the solar system, which are **the Roche limit, the Lagrangian points, the Titius Bode law and the Coanda effect** than to admit to Newton's failings.

They would not admit to Newton's failings because then the entire world will see they know less about science than does a pig know about history. They would rather put the error on the solar system than they would commit to the blatant mathematical cheating that Newton committed. It is the Universe that is always at fault when Newton becomes incorrect because without Newton's fabrication of science they have nothing to show for all the wisdom they try to pretend they have. Newton fabricated "Kepler's laws" has some correctness mixed largely with a farce and blending the truth with total fabrication of reality hides the lie behind something presenting the truth.

I did envisage a great deal of resistance after I came to realise the level of dishonesty in the ranks of the physicists but I never could imagine the fear there was in the ranks of publishers. Read the evidence I provide in the next few pages and see how undeniable their corruption is and how simple the corruption is to detect. Yet when I send out a book to publishers I send it to eighty on one day. Never is there even one publisher or academic willing to read my work never mind get around to publish it. Those physicists apparently put the Holy fear of God into the publishers! Now se the evidence and then you arrive at your conclusion about who is correct, science or nature because I play no part in the equation.

This is the solar system. Can you the reader see that the smallest planet is on the very outside and the second smallest planet is on the very inside? Can you see the two largest structures called planets are in the very middle? Notwithstanding planets or dwarf planets that is making dust to hide the truth! I am not only fighting a myth but I am fighting the brainwashing we all had to endure to force us to believe the myth they call Newtonian science. Whatever I say can't penetrate the layers of psychological abuse students went through and the mind control they were submitted to in order to believe it is mass that pulls any person and it is mass that creates gravity. The person that believes mass is the factor that positions planets never saw the solar system as it is and any person with a clear mind that sees the solar system would not be able to fit mass as a factor of allocating structures into it. Between Mars and Jupiter there is a band of debris we named the asteroid belt. It is a lot of rocks that remained in a group after Jupiter destroyed the planet by implementing the Roche limit because this planet at one time was exceeding the Lagrangian points.

Please look at the picture. While looking at the picture tell me how is it possible that the "mass" of Mercury could put it in the place it is and the "mass: of Jupiter could put in the centre where both hold positions according to "mass". I have been saying this so many times I am getting sick of repeating it but this formula they say positions planets. They say or then Newton said and they confirm that planets are in space in accordance with the formula $4\pi^2 a^3 = P^2 G(M + m)$. Now it happens just what they want to happen because now you are scared witless. When looking at this you are intimidated beyond your senses and you feel so stupid your inferiority takes control of your thinking and the feeling of utter stupidity wants to make you run and hide. This was the scare tactic Newton put on everyone he came across. No one dared ask him a question because the man was bewildering intellectual. That is bullshit because I as stupid as I am can look at this and see right through his menacing bluff.

This says $4\pi^2 a^3 = P^2 G(M + m)$ that the circle ($4\pi^2$) formed in the three dimensional space of the orbit of the planets (a^3) is equal or is the same as the location in which the orbit is () and this is determined by the gravitational constant (G) directly in measure wit the mass of the sun and the mass of the planet (($M + m$)). Go on and put in the mass of the sun and the mass of Jupiter and show how this can locate the position the planet is as they try to prove the formula $\left(\dfrac{P}{2\pi}\right)^2 = \dfrac{a^3}{G(M + m)},$ proves. You can read these formula as if it says @#%$&^* and it will have the very same meaning because it is completely invalid. ...And if that won't scare you socks off you they put you in a mental torture chamber by throwing in the formula $\left(\dfrac{P}{2\pi}\right)^2 = \left(\dfrac{a^2\sqrt{1 - \varepsilon^2}}{\ell}\right)^2 = \dfrac{a^4(1 - \varepsilon^2)}{\ell^2} = \dfrac{a^4(1 - \varepsilon^2)}{a(1 - \varepsilon^2)GM} = \dfrac{a^3}{GM}$ and one look at that will make you run and hide behind your mother hoping she will get these clever physicists away from you! Think how this formula will have any first year student scared witless and he or she would be great full NOT to apply it but just to learn this off by heart and repeat it one million times to secure the validity thereof as the truth. This is fabricated lies big enough to be worthy of any politicians commitment to honesty and if any physicist or Astrophysicist want to disprove what I say use it and show everyone how big a liar am I and how trustworthy Newton and his mock mathematics is.

Let's look at what the cosmos uses instead of the mass as Newton said is there.

The presenting of the cosmos in this picture is used to falsify the truth. They get away with it by saying the solar system can't be picture perfect presented in the ratio that it truly is but they never say this when showing a picture. The truth is that a planet doubles the distance every time a new position is allocated.

Mars is as far from the earth as the earth is from the sun as is from the sun.

The asteroid belt is a far from Mars as Mars is from the sun.

Jupiter is as far from the asteroid belt as the asteroid belt is from the sun.

Saturn is as far from Jupiter as Jupiter is from the sun, not from Mars but from the sun.

Uranus is as far from Saturn as Saturn is from the sun.

Neptune is as far from Uranus as Uranus is from the sun.

How can I put it more vivid than what I just did and yet I can't break through the brainwashing science put on people. Newton is not there…it is the Titius Bode law that is there in place in the solar system. If you still wish to accept Newton then you are too damaged by brainwashing to repair your mental status. You can accept Newton and remain part of the hoax or you can accept not me but nature. In the cosmos the reference point in doubling the distance is the first inner planet and because Newtonian science is incapable of producing any type of explanation they cheat like the mafia would by

telling everyone Newton's formulas $$\left(\frac{P}{2\pi}\right)^2 = \left(\frac{a^2\sqrt{1-\varepsilon^2}}{\ell}\right)^2 = \frac{a^4(1-\varepsilon^2)}{\ell^2} = \frac{a^4(1-\varepsilon^2)}{a(1-\varepsilon^2)GM} = \frac{a^3}{GM}.$$

This phenomenon I just described is named as the Titius Bode law

Most persons reading this has never heard of this but it is what truly is in place in the solar system when one study the layout of the solar system. There is no place for mass and you can see how I present the mass of every planet as you read later on. Pick an argument about this and I must refer you to the cosmos because I only show what is in the cosmos. Now we get to the point where I reveal what it that I discovered is. I discovered how the Titius Bode law in principle works. The Titius Bode law is the way that the Universe grow and it is how plants position in accordance with a precise ratio that is so accurate one can put values in a cyclic formula as follows.

$P_n = P_o A^N$

P_n = **period of orbit of the nth planet**

P_O = **period of the sun's rotation**

A = semi major axis of the orbit.

This is what the cosmos uses so if you have a problem with me condemning the Newtonian version take it up with the cosmos because I only show what is used by the cosmos and that which the cosmos uses is not what Newton said. So you have the dubious honour to stick with Newton and discount what the cosmos uses or you can side with me when I explain how this layout forms.

Should you now wish to read more about the Titius Bode law go to page 128

I am not going to reveal why it uses this system because that information the reader must purchase when they purchase the books I offer in the partnership. That information has never been known for as long as man is on earth and knowing that took me a life-time of study to accomplish.

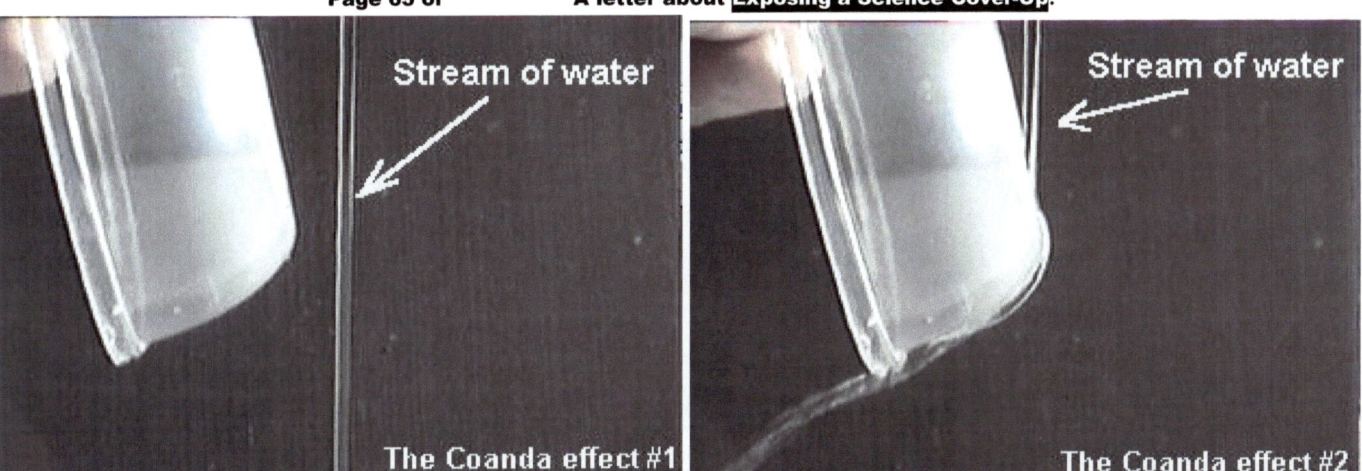

Stream of water

The Coanda effect #1
JL Naudin - 09-26-99

Stream of water

The Coanda effect #2
JL Naudin - 09-26-99

What this is goes by the name of the Coanda effect. I prove that this represents gravity as a principle more than any other form or factor could. Yet, with this so prominent in physics, you will never see any explanation about the Coanda effect in any physics handbook because the Coanda effect puts a serious question mark behind Newton's idea about physics allowing this.

The Coanda effect is the very reason why the Earth has an atmosphere, but you will not learn anything about the Coanda effect… because with the limited view that science at present portray they have no explanation for the Coanda effect or the atmosphere being there other than the mass of the atmosphere pulls the atmosphere down.

What a lot of unproven Newtonian gargle that is; what mass could the atmosphere have? By going to Lulu.com and then downloading **The Absolute Relevancy of Singularity The Website** http://www.lulu.com/content/e-book/the-absolute-relevancy-of-singularity-the-website/7517996] you will see how the spin of the earth compresses the space and by compressing with movement, not with mass pulling, the turning produces the Coanda effect and the Coanda effect by gravitational motion condenses space to become the atmosphere.

The Coanda effect is the principle that also proves the sun is not a gas giant but it is a liquid as can be seen from the liquid spewing from the sun's surface. That is a dead give away about what gravity really is. The earth spinning contracts the atmospheric space surrounding the earth and that process cause gravity to attract and not the mass of objects as Newton insisted. It is the space holding the object falling that moves downwards and not the object that falls. That is why Galileo was correct when he said all things fall equal under the same conditions notwithstanding size differences.

What Newton says is that things fall by the value of mass bringing on gravity. This means if everything has a different mass, therefore everything must fall at a different pace, which doesn't happen…and in that is where science is making the biggest mistake. I ask any one in science to prove the fact of mass applying, not as weight but as a force. Read on and I will show you how those in physical science are brainwashing students into believe that mass is responsible for gravity as a force.

I discovered how the Lagrangian points in principle works. The Titius Bode law is the way that the Universe grows and it is how plants position in accordance with a precise ratio that is so accurate one can put values in a cyclic formula. These are the points used by moons and safelights in orbit as they circle around planets. Again I repeat, those with a problem about Newton not working in science you better go to the cosmos and find out why the cosmos rejects Newton as much as I do because I follow the cosmos.

I presume Newton was completely unaware of the rings around Saturn and the other planets but these rings alone put Newton's claim on mass pulling mass in serious doubt.

This one of the four of the pillars I introduce by which the Universe works. You can say I introduced it because the Universe introduced these four comic principles when the first spot exploded into a dot. The four works together and never can they be separate. However in certain conditions in movement one might stand out more prominent in the relevancy in which the phenomenon is at that very place in that space. The phenomena are there used by the cosmos but science never understood why the phenomena are there and so science not wishing to show how stupid they are acted even more stupid and ignored the phenomena altogether.

I wish to quickly show one example of the mathematics by which these phenomena is calculated. A sphere has a line forming the axis. That is three points. The circle by which the spin runs holds four relative positions and that brings four. When moving this lot goes square because the points displace space and locate new positions. Therefore there are 3 going square = 9 and four going square = 16. Gravity is Pythagoras so according to the law of Pythagoras we have 9 + 16 = 25 and the root of 25 is 5. Therefore the next point to become part of the Universe when the Universe started in singularity is five and that position is 5. Everything is so incredibly simple if the correct lines in thoughts are used.

The stars show the Roche factor as the result when the Roche limit is bridged. The Roche limit is one of the four pillars I explain. Newton could never come close to even guessing why the rings are forming and why the star acts in the manner it does and as we see it does. You either side with Newton and know nothing that is reality in the Universe or understand the Universe and know why the Universe uses what is there that applies as physics. The gravity left behind as circles is Π or space. Gravity isn't mass. Gravity is the forming of space as the factor Π. Therefore gravity is not mass related but it is Π and the circle that's prominent is gravity that remains behind as Π in space. If you disagree with me take it up with God.

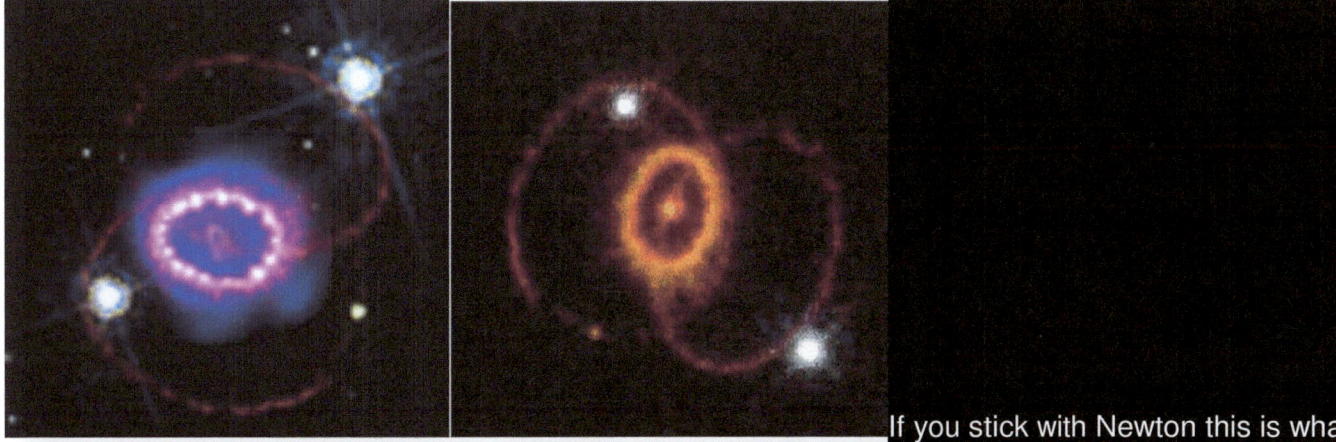

If you stick with Newton this is what

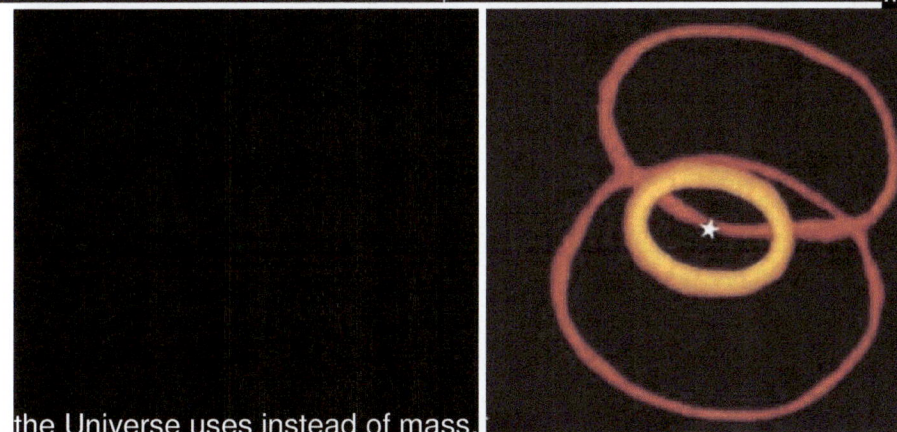
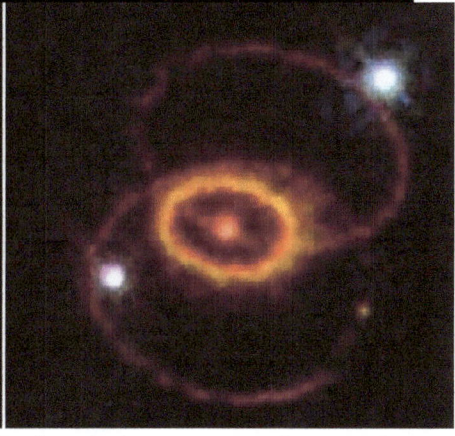

the Universe uses instead of mass.

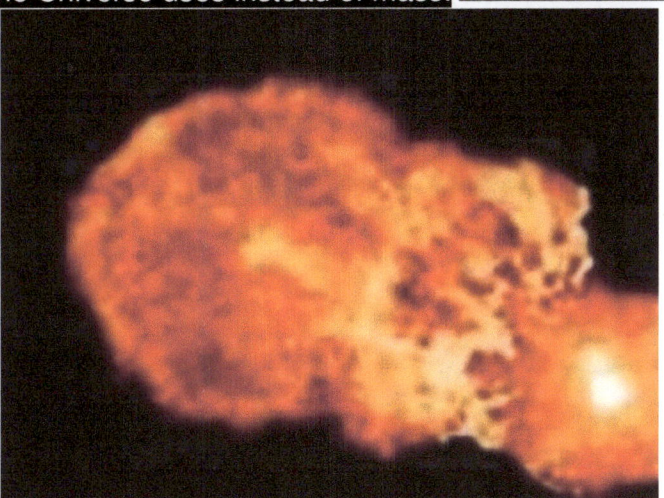

The Roche limit is the region surrounding each star in a binary system, within which any material is gravitationally bound to that particular star. The boundary of the Roche lobes is an equipotent surface, and the lobes touch at the inner Lagrangian point, L_1, through which mass transfer may occur if one of the components expands to fill its lobe. This puts the Lagrangian points system in the very heart of the Roche limit and made me realise that all these four phenomena are interlinked and moreover interwoven. It names after the French mathematician Edouard Albert Roche (1820-83). This means that since the 19th century Newtonian science was aware of the Roche phenomena but never took any effort to try to explain this because this kills Newton's mass idea off more than most other cosmic phenomena. This phenomenon prevents stars from colliding and that totally jeopardises the entire idea of Newton and his mass. I remember thinking that the behaviour of the stars are connected by some cosmic connection that requires urgent research. Then I started dipping into cosmology with unrest and this is how it started if my research started at any given point. This made me realise gravity is something to do with Π.

Everything in these pictures is heat and it clearly shows that heat is space gravity compressed by applying Π. Everything there is in these pictures and even the blowout happening is a circle applying gravity as Π. How can any one that pretends to be serious or professional in cosmology or even interested about science miss it? This started me off looking for something better than was in use.

There are two concepts involving the Roche phenomenon. The one is the Roche limit and the other is the Roche lobe, which is what happens when material move faster through space than the applying relevancy

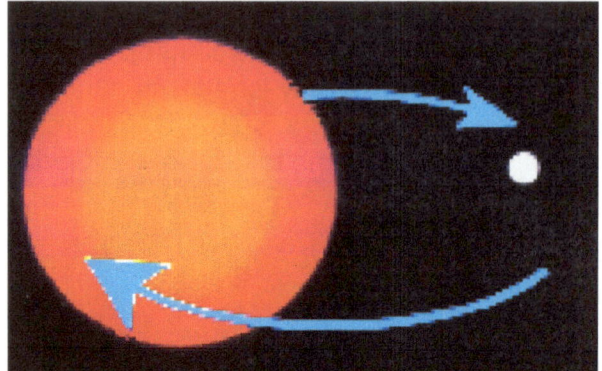

would allow. When I studied this I came to the conclusion that there is no mass pulling mass. The entree Universe must work another basis that what science propagates.

The system work as follow: There are two or more stars involved but lets keep it simple and only explain a "big" star and there are a "small" star and the "small" star is circling the "big" star just like the moon circle the earth.

For the Roche limit to become effective the "small" star must be with in the radius of 2.4674 the diameter of the "big" star. This puts the "small" star within the atmosphere of the "big" star.

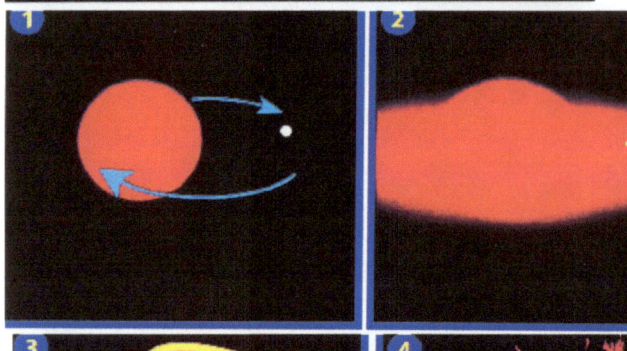

Then the "big" star starts to resolve the component forming the "small" star. The "big" star starts to spin the "small' star and this helps to convert the "small: star to become the density of the space surrounding the "big" star.

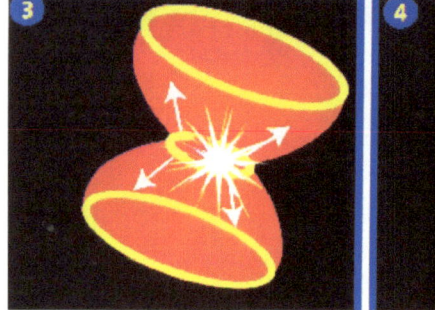
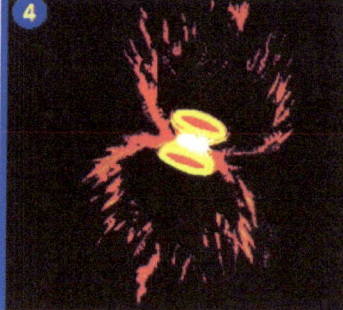

Because of the difference in movement the density of the "small" star is much less than the density of the "big" star. Therefore the "big" star vaporises the material forming the "small" star to the same status as the atmosphere of the "big" star is.

Since gravity is the contraction of space and the "small" star becomes the liquid atmosphere of the "big" star this then forms the space within the "big" or forms what we see as the atmosphere. The "sound barrier" is the Roche limit applying.

Coming to the conclusion about gravity being motion and mass being the restriction of motion was the easy part. The facts that presents the understanding of what produces the motion and what prevents the restriction from overcoming the motion was the part that required thinking. Figuring out why was everything on the move and where did the motion stop, well that was the part that took some figuring and some explaining.

What makes gravity move and why does gravity move…the answers are in the four phenomena never yet explained to satisfaction but now turns out to be the cradle of gravity. The answer can only come when the full content of gravity is fully understood as being the unexplained phenomena that produce in conjunction with one another the totality of gravity as we experience it.

They are the following:
Gravity is The **Roche limit,**

Gravity is The **Lagrangian system**

Gravity is The **Titius Bode law**

Gravity is The Coanda affect

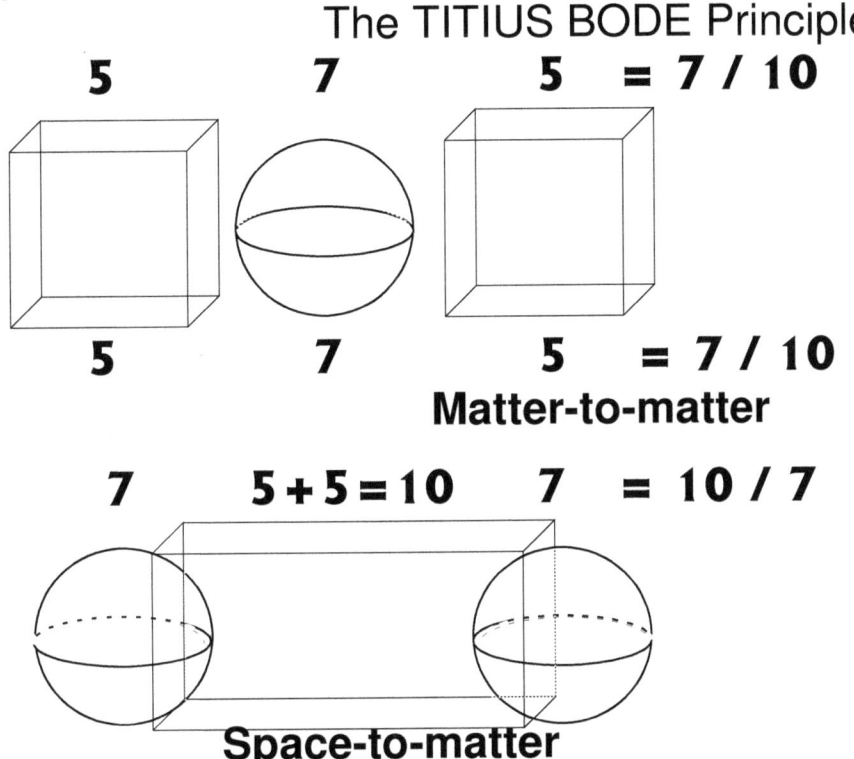

The TITIUS BODE Principle

5   7   5   = 7 / 10

5   7   5   = 7 / 10

Matter-to-matter

7   5 + 5 = 10   7   = 10 / 7

Space-to-matter

In the Roche limit the space factor provides occupied space-time and therefore the value of r is replaced by the value of Π bringing about a square in half of Π.

Gravity is the dimensional changing of heat holding r as reference to the sphere holding Π as the reference. Heat occupying space has the cube that can apply r, as a straight line bringing about the cube with all its other names than may find attachment to specific form but nevertheless still remains only a six-sided cube with angle changing in some cases.

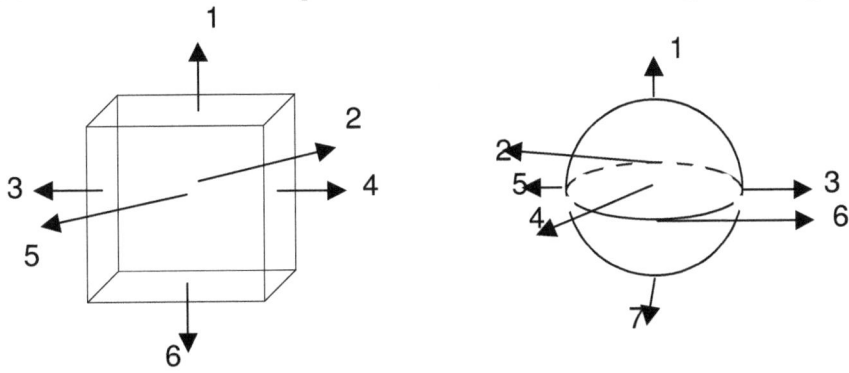

In the sphere there is no radius but only the extending of **k** from the centre **k** in six opposing directions relating to one another by the square but remaining Π because of the unity the matter holds in relating to space. In every sphere there then are the seven Π relating in precise dimensional and positional equality to the centre Π as well as to one another by 90^0 and 180^0 implicating the dimensional positioning.

Therefore the sphere holds **7** points and the cube holds 6 sides.
This puts material as the eternal sphere in a relation with space that without bonding forms the everlasting cube. The material as a sphere performs movement and the space as a cube allows the movement to take place.

The diagram at the top shows a cube and sphere with numbered arrows 1–5 (cube) and 1–7 (sphere).

By coming into contact with the sphere the cube loses on dimension to the seven dimensions dominating six bringing about that the cube then has 5 sides to the seven of the cube. That is the Lagrangian system with five cosmic atoms holding relevancy to the centre cosmic atom where the centre cosmic atom stands in for seven and the orbiting cosmic atoms standing in for five positions in space. There is a more explicate explanation about this somewhere else in this book.

The Titius Bode law is an extending dynamic deriving the law from the gravity dimensional factor where the space factor in a square of ten relates to a matter factor in the square by half (half since nothing can be in two places in the universe simultaneously) of the matter factor of Π or the square of space (10) relate to the matter factor of 7

The Roche limit is:

The region surrounding each star in a binary system, within which any material is gravitationally bound to that particular star. The boundary of the Roche lobes is an equipotential surface, and the lobes touch at the inner Lagrangian point, L_1, through which mass transfer may occur if one of the components expands to fill its lobe. It names after the French mathematician Edouard Albert Roche (1820-83).

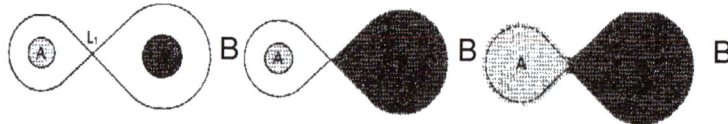

THE ROCHE LOBE: In a binary system, the Roche lobes of components A and B meet at the L_1 Lagrangian point. (a) In a detached system, neither star fills its Roche lobe. (b) In a semidetached system, one massive component, B, fills its Roche lobe. (c) In a contact binary, both components overfill their Roche lobes and share a common envelope.

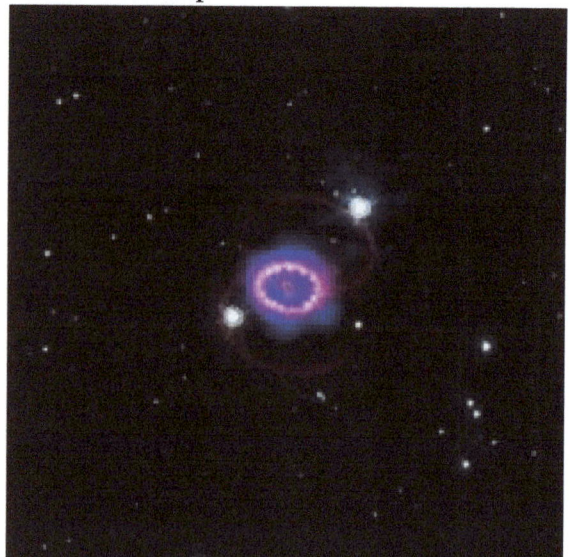

This is an example of the Roche lobe where the star divides into sectors all keeping $\Pi\Pi^2$ as format and yet it distributes its entire content while still maintaining $\Pi\Pi^2$ as format.

The star did not go mad but in the law governing the Roche limit it overheated and it then could no longer maintain structural integrity and therefore it expanded as it compromised gravity or movement.

It is all about controlling the heat within the star as to keep the structure un-compromised by creating sufficient gravity.

The entire Universe works on a balance between hot and cold, solid, liquid and gas, movement and density established by varying density related to movement between occupied space and unoccupied space.

When the star overheats it expands just as any container of fluid does when the liquid turns to gas. When the water in a car's radiator overheats and turns to steam the radiator expand and blow to bits. That is the nature of nature and that is the most powerful force within nature. By overheating the unlimited cold the Universe parted that which is cold and can't move from that which is hot and can never stop moving. This science think of in terms of naming it the Big Bang but that only represents one of so many stages through which the cosmos developed.

The ROCHE LIMIT

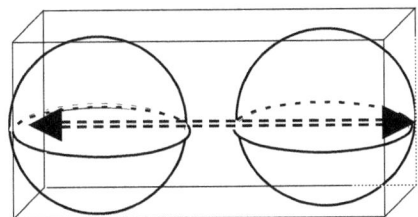

$$5/2 = (\Pi / 2 \times \Pi / 2) = 2.4674$$

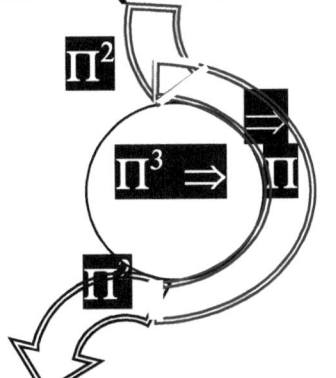

The COANDA AFFECT

$$\Pi^3 \times \Pi = \Pi^2 \times \Pi^2$$
$$= (\Pi^2 + \Pi^2) \text{ FIRST FORMING}$$
$$(\Pi^3 \times \Pi) \text{ SECOND FORMING}$$

Without resolving the concept that these four comic pillars present I would never have been able to figure out how the cosmos works, why the cosmos works and how everything within the cosmos fits together.

I call science a scam and if many readers if not most will completely disagree with me about science living as a scam then the next medical scam I wish to bring to every one's attention will support me. This scam is called humanity as it involves the medical science more than physics. It covers not what you are told or what you are taught but what science remains silent about.

It is what is never said that is of serious importance. It is never what the research reveals but what science hides from getting revealed. It lurks in the profits of the pharmaceutical companies or the doctors or the hospitals the industry never whish to reveal. Euthanasia is not about the absolute sanctity of life but it is keeping the patient alive when the patient is going to need the most expensive drugs and the most expensive intensive care. I saw a patient with a knife in his head wheeled out of hospital to find some other place to die because that patient did not have medical care to fund his desperate situation.

Because there was no money for treatment there was no treatment and this goes on in every private hospital in every city around the world. If that person had a good medical care and the medical insurer was willing to pay then a team of doctors would fight day and night to keep the person alive and no cost will be spared. It is during the last few months / weeks/ days / hours that most care and medicine is needed. During that time when the end of the life approaches and the fatal disease is going to take its final toll that the pharmaceutical company, the hospital and its staff, and every person sanctioned with keeping this individual alive will fill the bill as fast and as hard as possible to get the last money from the dying sucker.

You have a lawyer walk in there holding your last testament that states you are no longer presuming responsibilities for the costs and you have the lawyer tell them if they don't guarantee success with the treatment payment will stop on that minute and the costs for keeping you alive will be on the hospital and its staff who treat you and then you see how quickly they stop machines keeping you breathing. If you start legalising euthanasia you kill the part that serves the highest profits and they would rather see you suffer than they would agree that you might die cheaply.

If you are not prepared to pay for their effort of holding your life ever so dearly they will lose all interest and let you die as quickly as the machines are switched off. Your life and all life have value when there is someone prepared to pay the bill and those doctors are the worst criminals there mathematically is. But yet again behind this medical industry there are insurers and behind the insurers there are bankers. Fighting euthanasia is a conspiracy because euthanasia will kill the huge profits the medical industry makes. If we are all so against euthanasia let's take the profits of these pharmaceutical companies, the hospitals and the doctors and spend their money to pay for every one dying that can't afford treatment. Now it is a case that they would rather see people suffer and die in agony when such persons can't pay for the treatment because after all being part of science makes them equal to God and therefore others must suffer so they can reap the reward of the money they invest in promoting and furthering science.

For decades I tried to come to terms with the inability there is in science to explain the cosmos in real terms, when using the science of official reputation. That which there is makes a mockery of science because the undisputable clues left in the cosmos makes what little correct explaining there is available, seem like a comedy of errors, when it is mixed in with all the other near Dark Age errors we still use after so many centuries that provided countless opportunities to revise the old muck. By applying current accepted Astronomy as such the phenomenon found all over the cosmos is still beyond the explaining ability of Mainstream science. This is true and it is a shame because it also is an undeniable fact in spite of the vast knowledge and progress in other forms of science taken in the manner science uses when it approaches cosmology.

Cosmology truly lagged behind while the understanding and advancing of physics, mathematics and chemistry as subjects were flourishing. By comparison I saw how little there was available in explaining cosmic phenomenon and how much improvement in understanding the other departments such as chemistry, electronics, medicine etc. could offer as results were coming about from research. Even where there is a little explaining available in cosmology it turns out that such explaining is confusing to say the least and at best it highlights the manner in which science is applying double standards. For decades photographs were the only progress forthcoming as an addition to improve the meagre field in cosmology and that improvement was artificially stimulating cosmology.

By providing a false impression of advancement, everyone missed what and how much was missing…To the connoisseur desperately looking for more than the obvious stirred in with some out-dated misinformation dating back to the Middle Ages, it all seemed as if it was a picture portraying the ridiculous to make the sublime look good. The pictures only proved the opposite of what progress in cosmology will represent. In truth and as such in cosmology the cover up that was hiding the lack of progress about the science of true cosmology was only forthcoming in the improving of electronic optical telescopic advances and spectroscopic progress.

There were only photographs carrying beautiful pictures which pleased the less informed except the photographs did not bring progress to cosmology at any intellectual level by promoting insight. The explaining that the photos demanded about the subject had the opposite effect of installing hope because what it did do was underline what lack in any notable progress there truly is in our understanding of cosmology and laws in the cosmos. Getting as far as realising the conspiracy took me down roads overgrown by ignorance and which I had to uncover myself as if hacking away miles of overgrowth with a machete chopper. All of the disbelief science showed to my work in the past and their refusal to see past Newton made any and all attempts on my part as bad as they could be, strangling and smothering my attempts to announce my uncovering of the newly found insight on my part.

Their mannerism in blocking and frustrating my opinion when showing the mistakes in science convinced me about `a Conspiracy in Science in Progress` and this spurred me on to tell the entire world about their brainwashing students minds. By the manner they selectively withhold information when teaching science, amounts to deliberate brainwashing of students in physics by "normal" education practises. The new concept I wish to introduce puts all emphasis on space ands material is only space filled with material substance while other space is filed with non-material.

In the end all space are equal but the movement it has makes the difference it presents in relevancy. All space structures hold in the centre most heat concentrated and from that centre holds all material owned by that structure. I can go on and on but heat in the centre couples gravity to space-time, just like Kepler said before he was spoken for on his behalf and without his permission or his agreeing to it.

Studying Kepler helped to understand why the phenomena are there to begin with and that enabled to explain in some way…

Why is the Universe depicted as a sphere…and why would that then be correct…
…how did everything become so much and so large…
…why did it start small…

The Titius Bode law
That I explain

…why does it grow from small to large…
…why was the start so small…
…why is it growing…
…where is it going while it is growing …
…why is it any specific size...
…what was everything before that…
and why in creation would this lot then reduce again!!!

T^2
Gravity is a^3 k

The Coanda Principle
That I explain

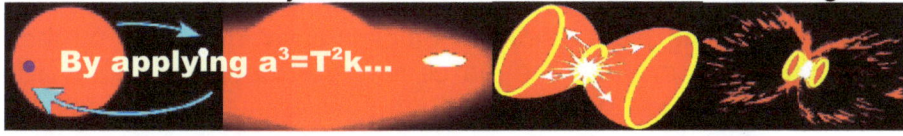

By applying $a^3 = T^2k$…

The Roche limit becomes self-explaining through the Coanda effect when using the Coanda principle in the practical

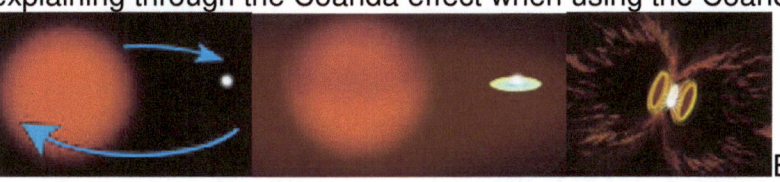

Every one in science throughout many centuries ignored Johannes Kepler because all saw him as some derogative of Newton…until now. Kepler introduced space –time but nobody took the time to acknowledge Kepler's introduction. Kepler introduced space a^3 – time T^2k and showed that it is space a^3 – time T^2k that is performing gravity by relevance of k.

Are our centuries long ignoring of Kepler truly the answer…Kepler introduced gravity by principle but no one in four hundred years took any notice of the manner in which Kepler brought gravity into human conception and understanding?

Kepler calculated that it is the motion of space a^3 during the time T^2k that forms the gravity that is keeping the sun and all the individual planets apart but moreover gravity is keeping the planets in orbit. While every one was surprised but now accepts there is a growth in the Universe by which the Universe is expanding…for four centuries Kepler said that and no one took notice.

According to Kepler the expanding is the normal trend that the cosmos will follow…that he said four hundred years ago…yet in spite of Kepler findings…science still clings to the idea that what keeps the Universe secure is contracting the force by the mass value that creates an attracting in the distance between the objects.

That is NOT what Kepler said.

Gravity is the effort of independent objects to secure their position as the centre of the Universe by motion of space in space in relation to space by moving through space.
Let this book reconsider your verdict about what gravity is, because by you reading it, it most probably will…

All Kepler's charts prove one thing and that is that space a^3 moves T^2k. Moreover, all space a^3 moves T^2 in relevancy of factor k. All space moves because the Universe is constantly changing form and formations as time forms and alters space. While moving outer space expands bringing about movement and material within combined structures such as stars contracts by spinning motion which is gravity, which is the opposite form of the expanding movement we find that outer space produces.
The pendulum proves my point that gravity is space moving down towards the earth and not "mass pulling mass". The pendulum proves that mass has nothing to do with gravity. Their pretentious mathematics becomes religiosity and this religious belief in the infallibility of their Godly mathematical insight becomes their raving stupidity. Their ignorance is that which science must hide. It is that stupidity which they hide, which is what you don't see. Science hides what science doesn't know which they cover under the larger pretence of their cleverness. Also science hides from everyone's view what

science doesn't know so that students will never realise Newtonians are covering their stupidity and ignorance behind a curtain of arrogance.

The pendulum is one example of them hiding the obvious because they can't see the obvious. This conspiracy to not reveal their stupidity then becomes the conspiracy they nurture for centuries. Science conspires to hide by covering their stupidity from open view. The main thing is that science is clueless about time and time is the driving cosmic plan so science hasn't a plan to see the truth. I have been in fruitless conversation mainly going one way about what I see as their stupidity in defending flawed science. The pendulum uses gravity to measure time and they can't see this science in over 300 years?

I told them this but I received no response. Still notwithstanding their arrogance about ignoring my showing right from wrong, still those physicists filling high academic office are so infatuated by their superiority and their personal righteousness they can only see the malice that the Pope showed towards Galileo when he differed from the general science views and yet they do the very same today. They are overwhelmed by their correctness.

You will see how those hypocrites condemning the pope having the holier than thou attitude because they think they are cleverer than all other intellectually lesser sub-humans. They accuse the Pope of constraining science progress back but still to the present they fail to award Galileo's vision with the correct attributions. At least the Pope allowed Galileo to print a book while they did everything to stifle any effort to show what the true science is behind the discovery of Galileo.

Whenever anyone shows those in science that their religiosity called Newtonian science is wrong, they act exactly like the Catholic Church did to Galileo five hundred years ago. Galileo showed all things fall equal and if that is true mass has no effect on falling but they condemn the Church for not excepting Galileo's principles while they cover-up. I show how and why the establishment of science frustrate me by furthering a cover-up.

The crude thing is that I prove all those filling academic office in science have also still not accepted Galileo to the letter and I challenge anyone to prove me wrong. That is one of the parts of the big conspiracy I see and call it the Mother Conspiracy! By the swinging pendulum in space while connecting it to the earth Galileo proved gravity is time and that principle goes unnoticed as it diverts from Newton's idea that "mass" produces a magical "force" of a "pulling nature" he called "gravity".

This view they can never explain while this is what they will always defend as the truth. Those oh, so clever Dark Aged wizards pronouncing the upholding of free speech and science liberty for all of mankind while hiding behind their superior positions will befall the same memory as what befell the Pope in the days of Galileo.

They will be remembered for being the last of those who were stupid enough to believe that an inexplicable magical "force" of "gravity" "pulled" the Universe by "mass" while E.P. Hubble showed the Universe is not contracting but otherwise expanding.

Their sublimation covering their stupidity causing their ignorance will outlast them for all time to come. As the Pope five hundred years ago is remembered for his ignorant stupidity, so those currently in office would also befall the same fate just because they were as stubbornly arrogant and ignorant as the Pope was back then when ignoring mistakes. I know one thing for sure and that is that no one in science will read this book because everyone in science do not wish to be confronted by what is the truth about science and therefore everything in science will hide behind a cloth that covers all.

Dan Brown brought in another lie faked as the truth and was awarded a billion dollars for forging a lie to rape the truth. I live in poverty because I wish to remove a lie from science and for that I live in poverty and I am ignored like a rabies dog. That is gratitude but also that is the reality of science living the lie faked as the truth. Let's investigate lies and truths and see what hides as deceptions in places where we invest truths.

The only way Dan Brown could be correct about his account and his entire saga is if Da Vinci was one of the members sitting at the table and drew from what he witnessed the night of the Last Supper but otherwise it is a fable that science silently underwrites for other motives.

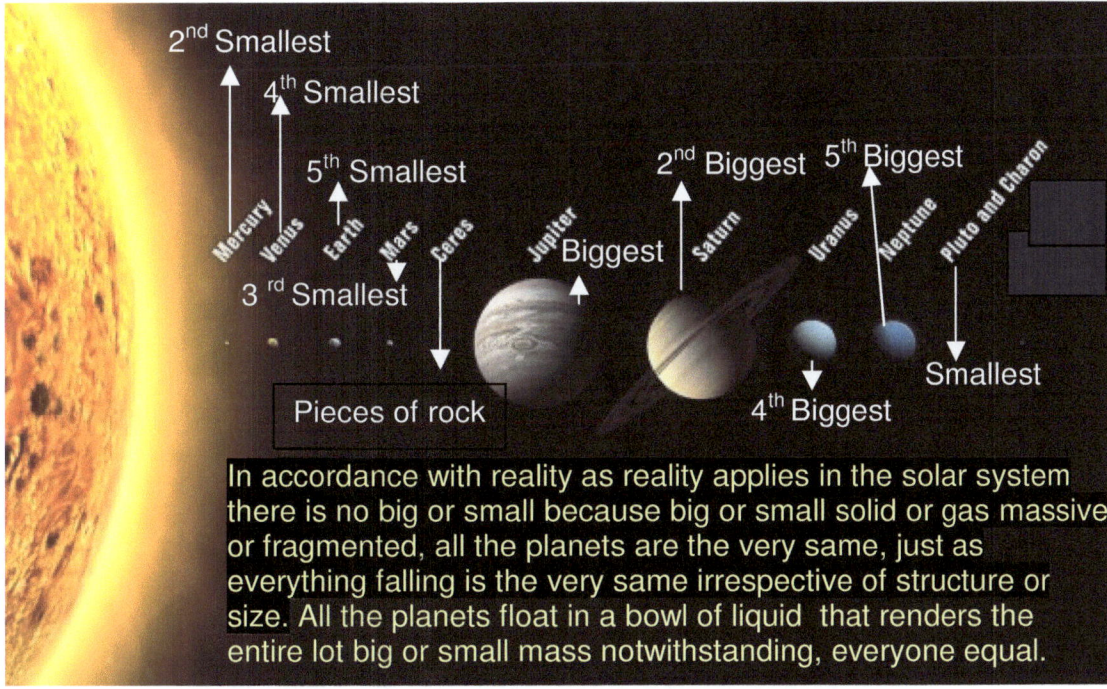

By depicting the solar system in such a presentation as Newtonians normally do such as the picture next this form of presenting the layout without providing correct spacing purposely corrupts the entire structure formation by which the solar system develops. It then purposely hides the essence that forms the solar system.

This is so typical Newtonian in every sense there is in science. Can anybody, even those with the mentality of students and his gang see that the planets are not arranged from the biggest that is most massive and then therefore should be closest to the sun and smallest way to the outside as they should if plants orbits P^2 was the result of size or mass $G(M + m)$ and mass has no place in the layout?

I argue that if it is the correct practise to use $T^2 = \dfrac{4\Pi^2}{G(M = mp)} a^3$ to calculate gravity then the positions should go according to the mass $\dfrac{1}{G(M = mp)}$ but by showing another table that Kepler devised it is clear that the way Newton changed the tables of Kepler just did not match the ideas Kepler introduced. Kepler introduced $a^3 = T^2k$ and by calculating the numerical values the total valuing k at a minus value is $T^2 \div a^3$

If Newton said $F = G \dfrac{M_1 M_2}{r^2}$ which is totally ridiculous and yet science tells nature this is what must be.

PLANET	Mean Distance from the Sun (AU)	Equatorial Radius (km)	Mass of planet (Earth=1)	Mean density (grams/centimeter³)	
Mercury	0.3871	2439	0.06	5.43	2nd smallest but closet to the sun
Venus	0.7233	6052	0.82	5.25	4th smallest and 2nd closest to the sun
Earth	1.000	6378	1.000	5.52	4th smallest and 3rd to the sun
Mars	1.524	3397	0.11	3.95	3rd smallest 4th closest to the sun
Jupiter	5.203	71490	317.89	1.33	The biggest with "most mass" and bang in the centre and position 5
Saturn	9.539	60268	95.18	0.69	The second biggest
Uranus	19.19	25559	14.53	1.29	The 4th biggest
Neptune	30.06	25269	17.14	1.64	The 3rd biggest
Pluto	39.48	1160	0.002	2.03	The smallest of the lot.

The Sun	**Mercury**	**Venus**	**Earth**	**Mars**	# Jupiter
	2ND smallest	4TH planets	5TH yet	3RD nearest to the sun	Largest planet

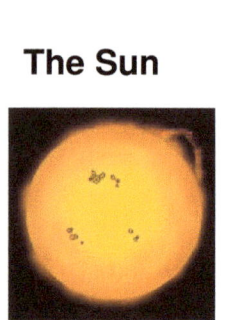

	.055 **earth mass**	.81 **earth mass**	1 **earth mass**	.107 **earth mass**	318 **earth mass**
Jupiter		**Saturn**		**Uranus**	**Neptune**	**Pluto**

Show me how "mass" plays any part in the   layout of the structure of the solar system?

 Prove the valid ness of    Newton's formula $\left(\dfrac{P}{2\pi}\right)^2 = \dfrac{a^3}{G(M+m)},$

318 **earth mass**   95 **earth mass**   17 **earth mass**   14.5 **earth mass**   0.002 **earth mass**

If "mass" placed the solar system into a position according to "mass" forming the constriction as Newton formulated it does with these "mass" values such as **Mercury** being **.055** times the earth mass, **Venus** being **0.81** times the earth mass, **Earth** being **1** times the earth mass, **Mars** being **0.107** times the earth mass and only then comes **Jupiter** with being **318** times the earth mass and still they insist "mass" rules the forming of the structure of the solar system.

On the other side of **Jupiter** this inconsequential order is representing the other planet layouts as the sizes are put in fractions of so many times the "mass" of the earth and yet as any one can see these values of "mass" completely ignores this fact of "mass" forming the positions of planets in the solar system as Newton insisted it does when considering the layout. Yet this is the fundamental basis on which physics in its entirety is based. Everything that physics rely on in order to exist depends on the correctness of this statement where "mass" form the allocations of the planets. Taking the layout configuration into account there is no evidence that the size or "mass" of any or all of the planets play any part in the structural composition or the allocation that the planets must have and makes a fool of Newton's formulas in that they pretend to have any ability to be used as mathematical calculations, as you will read in my work.

In this study of the layout of the solar system the last thing we encounter is "mass" since as Kepler's table shows all planets orbit the sun at just about an equal pace notwithstanding the incomparable size differences and the inconsequential orbit range of positions they take on. Look at the table of the orbit T^2 in relation to the area in which it spins a^3! The problem is that when looking at Kepler's table then if there is $T^2 \div a^3$ according to the table matching a column, then mathematically $T^2 \div a^3$ must be k^{-1} and where k^{-1} goes negative it shows space reduces time. It shows space in volume goes single by movement of space and not objects. The "Newton's mythology" comes from the fact that students have to learn what the professors claim to be true and what was never was proven. Students have to repeat in examinations that

the formula $F = G\dfrac{M_1M_2}{r^2}$ is truthful and viable while it was never proven.

Do you realise that it is an accepted practise that all students that are studying physics on all levels are subjected to the most intense brainwashing and thought control found any where on Earth? This must be some sort of a joke you may think but thinking that way in disbelief is just what those practising the mind control wish you to think!

We all were mentally processed to believe in science as much as believe science and what is said in the name of science and I will prove this statement. I prove there is or never was anything remotely, which they in science call "mass". At this point you immediately question my mental balance. Immediately you think in terms of "mass" as that which pushes you onto the earth and it is in that confusion that they rely to keep everyone dim witted and out of focus.

If I say prove the factor such as "mass" you think of that thing that pushes you down to the ground but that is not what they say "mass" is. They say "mass" is the contraction of heavenly bodies having some force of attraction on each other and as that there isn't such a thing. But they find it rewarding to have everyone confused between having weight and having "mass". When it suits them the two are the same and when it does not suit them the two are not remotely similar. Here and now I present you with the layout of the solar system and ask you to present "mass" as a factor that forces the solar system to adhere to "mass" pulling the solar system in a structural composition.

If you are not a person for figure then following the next argument in mathematical detail is not that important but you are most welcome to take this what I say to any person with a mathematical back ground and tests the truthfulness about my argument and how I show how simple it is to wash Newton away. In plane language to say $a^3 = T^2$ as Newton did is to say you being part of the third dimension can walk into a mirror of your image and hide there for a while because the third dimension or a^3 is the same as standing inside a flat glass mirror T^2 that has two ^{2} dimensions compared to your normal ^{3} dimensions.

In order to confuse and to mislead students starting to learn the most fundamental physics they bring an array of possibilities to the table that they present as correct and undisputed which was tested in the past and no one was ever able to tarnish the correctness of the facts even in the slightest because what they present is without doubt most correct.

Planet	Time T^2	Divided by or made relevant to	Space a^3	$k^{-1} = \dfrac{T^2}{a^3}$	k^{-1}
Mercury	0.058		0.059	$T^2 \div a^3$	$k^{-1} = 0.983$
Venus	0.378		0.381	$T^2 \div a^3$	$k^{-1} = 0.992$
Earth	1.000		1.000	$T^2 \div a^3$	$k^{-1} = 1.000$
Mars	3.54		3.54	$T^2 \div a^3$	$k^{-1} = 1.000$
Jupiter	140.66		140.6	$T^2 \div a^3$	$k^{-1} = 1.000$
Saturn	867.9		868.25	$T^2 \div a^3$	$k^{-1} = 0.999$
Uranus	7069		7067	$T^2 \div a^3$	$k^{-1} = 1.000$
Neptune	27159		27189	$T^2 \div a^3$	$k^{-1} = 0.999$
Pluto	61703		61443	$T^2 \div a^3$	$k^{-1} = 1.004$

They say Newton proved that $a^3 = T^2$ when he investigated the work of Kepler. This is how they mesmerise one and all because when any one investigate Kepler in terms of what Kepler said Kepler had three columns that had numbers and Kepler gave a formula by which THE COMOS no less presented facts to prove cosmic the movement forms as **Space a^3 = Time T^2k** or if singularity becomes part of the equation $k^0 = \dfrac{kT^2}{a^3}$ and towards material such as Kepler saw the Universe indicated movement as

$k^{-1} = \dfrac{T^2}{a^3}$. If Kepler calculated $a^3 = T^2k$ and we know mathematically this implies that **Space a^3 = Time T^2k** and we now know that **Space a^3** is a dimensional concept and where putting **Time T^2** in a square relation to distance **k** then time moves space by **k** point or to point **k** whichever you prefer. Then **Space a^3** must form by **Time T^2** moving and **depositing distance k.** Moreover centuries past since the time of Newton and every person that studied Newton or taught physics according to Newton must have seen there are three columns and if you divide the value in the column a^3 into the value in the column marked T^2 a different value other than 0 came about. Yet in all the centuries no one that played a role in physics was bothered to come clean about this misconception! The space in which planets are, is not 0 or nothing. By having **Space a^3** going singular or becoming one in relation to **Time T^2** and this puts the distance k^{-1} it means the space in **Space a^3** diminishes or compacts or become less in relevancy to what it was before. This is extremely important to remember because I will show that the **Titius Bode law** indicates material does not move towards the sun but maintains in a specific ratio.

Please consolidate this $P = \left(\dfrac{4\pi^2 a^3}{G(M+m)} \right)^{0.5}$ which Newtonian wisdom says is true with $P_n = P_0 A^N$, which is what the solar system have in place and which is what the solar system upholds.

There is no room in a room to show this layout in its full compliment where it covers all the nine planets. If mass formed gravity, then the layout should be running from the biggest to the smallest. The distance increases in relation to a ratio code and even in the case of debris called Asteroids keep in a predetermined circle in an allocated position in accordance with the Titius Bode configuration all though it is no more than rocks and even up to dust particles. These cosmic dust and debris are small and yet although not structurally up to any "mass" they orbit the sun just in the same manner, as does Jupiter.

This says the distance should be in terms of size $F = G \dfrac{M_1 M_2}{r^2}$, but it is not, it is according to the Titius Bode law, which is some law no one ever hears of because it disproves Newton and his mass concept. It shows Newton has no ground on which to form his concept that is completely wrong! But while the cosmos disproves Newton, science believes Newton in spite of the cosmos using the Titius Bode law. This is very typical of science in the way Science prefers to cheat the truth to prove Newton correct.

It is not "mass" that forms the solar system but it is a process-principle that goes by the name of the Titius Bode law. What forms the solar system is not mass but this is what is there: The Titius Bode Law Bode's Law" or "Titius-Bode Law". The original formulation was a = (n + 4) / 10 where n = 1, 3, 6, 12 24, 48... The modern formulation is that the mean distance a of the planet from the Sun is, in astronomical units (AU$_{earth}$ = 147.597 $*10^6$ km): a = 0.4 + 0.3 x k where "k'= 0,1,2,4,8,16,32,64,128 (sequence of powers of two and 0). The following table compares the law's predictions with the actual distances, where the addition of Pluto is a modern modification.

Planet	n	Titius-Bode Law	Semi-Major Axis
Mercury		0.40	0.39
Venus	0	0.70	0.72
Earth	1	1.00	1.00
Mars	2	1.60	1.52
asteroid belt	3	2.80	2.8
Jupiter	4	5.20	5.20
Saturn	5	10.0	9.54
Uranus	6	19.6	19.2
Neptune	-	-	30.1
Pluto	7	38.8	39.4

$P_n = P_0 A^N$
P_n = period of orbit of the nth planet
P_O = period of the sun's rotation
A = semi major axis of the orbit
This table shows the arrangement when the strictest of layout discipline is followed that adheres to the precise layout of the Titius Bode law and how the solar layout set-up forms. This is what there is in the solar system. This is how the solar system is built and therefore also the Universe. This is in the place of where Newtonians say is that it is "mass" that is driving the solar system. I say prove "mass" because I prove why it is the Titius Bode law that forms the solar system. The Titius Bode law is gravity because the Titius Bode law is forming the concept that I prove is Π and Π is gravity. Since I prove the Titius Bode law forms Π as gravity and not "mass", no one this far took the time to see what I prove all the while what I prove is the precise reason why the solar system uses the Titius Bode law to form the planet layout.

This is all explaining they with all their "Godly" wisdom could inspire and that is why no one ever heard of the Titius Bode law. I could mathematically figure out this formation but if they agree to my work and promote my work their work and all the work of those that came before them goes to science fiction where it belongs.

The Titius Bode law relates the mean distances of the planets from the sun to a simple mathematic progression of numbers.

To find the mean distances of the planets, beginning with the following simple sequence of numbers:
0 3 6 12 24 48 96 192 384

With the exception of the first two, the others are simply twice the value of the preceding number.

Add 4 to each number:
4 7 10 16 28 52 100 196 388
Then divide by 10:
0.4 0.7 1.0 1.6 2.8 5.2 10.0 19.6 38.8
The resulting sequence is very close to the distribution of mean distances of the planets from the Sun:

Body	Actual distance (A.U.)	Bode's Law <A.U.)< td>
Mercury	0.39	0.4
Venus	0.72	0.7
Earth	1.00	1.0
Mars	1.52	1.6
Asteroids		2.8
Jupiter	5.20	5.2
Saturn	9.54	10.0
Uranus	19.19	19.6

In twelve articles I explain how the four Cosmic Pillars form gravity but no publisher wants to publish it because the conspiring cheats wants to uphold Newton's corruption because the corruption is all they can understand. I sent it to many physics magazines and Universities and when I received an answer it was to insult me or otherwise I never even got an acknowledgement about receiving such a document.

…But they would not even think of publishing my articles because my articles prove that Newton's views are hogwash and corruption and where I show an answer to their madness they will not read my work because my work condemns their science completely. In their approach as they condemn my work is because what is not there in my work is Newton and what is there in my work and what they say should be there while they say is not there is Newton and because my work doesn't hail Newton they reject my work. If they keep on rejecting the truth I promote they keep the conspiracy alive that will cover-up my work and therefore they will never have to admit in their disinformation on which they and all their generations of predecessors built a hoax more elaborate and convincing than any religion before or since could ever master.

This is the best that the Newtonian can do with the most important principle forming one in four part of the entire cosmic code. The cosmos forms by code and I have unravelled that code. This is how gravity applies space by implementing the Law of Pythagoras and by the law of Pythagoras time builds space.

I am placing this configuration that applies in the order it should be and in the process I use small pictures to show the figuration in place.

If Newton is correct then ask any physics professor to explain:

Why does the distance from the sun  to Mercury double to Venus and that again doubles in

distance to the earth and that distance again doubles all the way to Mars and this carries on

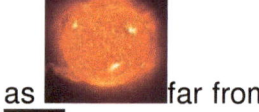 going throughout the solar system. Tell your physics professor to make sure he or she uses Newton and his idea that the mass every planet has forms gravity and to use only Newtonian gravity principles to explain this.

If you follow the tour and go where I guide you to go you will learn why in the solar system is Mars twice as 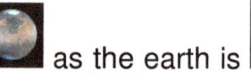 far from the sun  as the earth is  and why the earth is twice as far as Venus is away from the sun and Venus is twice as far as Mercury. This principle is called the

Titius Bode law and is in place in the 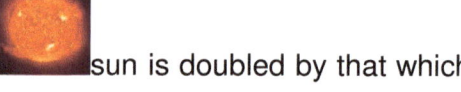 solar system instead of the mass factor Newton said is in use by the cosmos.

Let's repeat this again…

The distance that Mercury has from the 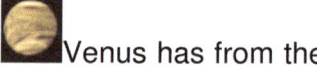 sun is doubled by that which Venus has from the sun

Then again the distance that 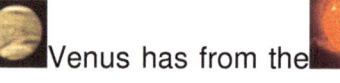 Venus has from the sun is doubled by that which  the earth has

Then again the distance that  the earth has from the  sun is doubled by that which

 Venus has and inexplicably this forms the layout of all planets in the solar system. But why does science never mention this? Professors in physics seldom mention the Titius Bode law when explaining how the solar system forms because the Titius Bode law makes rubbish of Newton's gravitational principles. It proves Newton is a fraud. This process forming distance between planets carries on throughout the solar system.

The problem mainstream science in astrophysics has is that science always tells the Universe what it should be instead of looking what the Universe is. It started with Newton telling the Universe about mass pulling mass and then studying Kepler's work making unfounded conclusions when not having a clue about the meaning behind the numbers of the work of Kepler, changing Kepler's work and that of Galileo to suit the ideas of Newton.

Kepler gave a table with numbers in three columns and still Newton saw it fit to ignore the one column completely and only grant two columns where Newton then proceeded to ignore the numerical values the Kepler received from the cosmos ad put $a^3 = T^2$. The concept formulated by Newton to present Kepler's work as $a^3=T^2$ is mind blowing ridiculous and the figures in Kepler's columns prove that $a^3=T^2k$. How does Newton find the audacity to destroy Kepler's figures? The man became power mad with fame and thought he could change the Universe and so he did. That is what science upholds to this day. He did the changes because he and everyone else presumed there is a number such as zero or nothing within mathematics and there can never be such a value.

The ratio forming allocated positions are the mean distance *a* of the planet from the Sun is, in astronomical units (AU_{earth} = 147.597 $*10^6$ km): **a = 0.4 + 0.3 x k** where "k'= 0,1,2,4,8,16,32,64,128 (sequence of powers of two *and* 0). The following table compares the law's predictions with the actual distances, where the addition of Pluto is a modern modification. This ratio is what applies and onto mass and this is what the cosmos uses in spite of and contradicting Newton's ideas. This puts Newton out as totally incorrect! This can' be that hard to understand because it is not mass that positions planets but the cosmos uses the ratio called the Titus Bode law. I prove why the cosmos uses the Titius Bode law but when I do I am criticized that I use too simple mathematics to be accepted.

PLANET	SEMIMAJOR AXIS $a\ (10^{10}\,m)$	PERIOD T (y)	T^2/a^3 $(10^{-34}\,y^2/m^3)$
Mercury	5.79	0.241	k^{-1} = 2.99
Venus	10.8	0.615	k^{-1} = 3.00
Earth	15.0	1.00	k^{-1} = 2.96
Mars	22.8	1.88	k^{-1} = 2.98
Jupiter	77.8	11.9	k^{-1} = 3.01
Saturn	143	29.5	k^{-1} = 2.98
Uranus	287	84.0	k^{-1} = 2.98
Neptune	450	165	k^{-1} = 2.99
Pluto	590	248	k^{-1} = 2.99

In the table Kepler concluded there are the following columns as "**a**" as "**T**" and as "**k**". In table Kepler presented the column under "**a**" has a value and "**T**" also has a value and "**a**" has a very different value to that of "**T**". What Newton neglected to say is when "**a³**" = "**T²**" which value does he work with? Does he work with "**a**" or does he apply the values under "**T**" because in the columns Kepler provide "**a**" is completely different to "**T**". So Newton started this habit of being smart and being so wise as to tell the Universe what it is and it is according to Newton "**a³**" = "**T²**" in spite of the cosmos giving Kepler numerical different values under each column "**a**" and "**T**". In the arranging

Mercury	$T^2 \div a^3$ = 0.983
Venus	$T^2 \div a^3$ = 0.992
Earth	$T^2 \div a^3$ = 1.000
Mars	$T^2 \div a^3$ = 1.000
Jupiter	$T^2 \div a^3$ = 1.000
Saturn	$T^2 \div a^3$ = 0.999
Uranus	$T^2 \div a^3$ = 1.000
Neptune	$T^2 \div a^3$ = 0.999
Pluto	$T^2 \div a^3$ = 1.004

Kepler's columns indicate a ratio and no ratio can result in zero!

Image Copyright JPL

I compiled **a new cosmic concept** by which I eliminated all the incorrectness that Newton has burdened science with but with this being my opinion I did not find a garage full of academics supporters waiting to applaud me and to uphold my views on the matter. Gravity rests on movement of material in relation to other material also moving. The movement of the sun provides the earth with movement but not only that al movement going straight becomes circular movement and circular movement takes place within the circle in which it moves going forward as a straight line and in that idea of a circle becoming a straight line and a straight line becoming a circle the entire concept of cosmic gravity is vested. According to the Big Bang theory the Universe expands and there is no evidence of pulling bringing about a Universe contracting or becoming smaller. The Hubble constant is sole evidence of this proof of expanding. Therefore I challenge the concept they build on the fact that mass attracts mass and everything is pulling everything else. Yet still I was not going to be ambushed by their relentless stonewalling my efforts and blocking my efforts in introducing both the incorrectness and the new cosmic theorem I concluded. My cosmic concept is that the Universe is about heat forming densities. It is the density of hydrogen making it a gas as much as it is the density of the massive Krypton, Xenon and Radon that makes these elements gasses although the gasses are the heaviest inert gasses and Lithium as a solid being the lightest solid. The gas I mention is many times over as massive as the lightest solid is and yet with all that massiveness, it is gas and gas floats in the air. The gasses form what they are because they are a mixture of heat and material putting a factor such as mass completely out of the picture. All materials are solids as they are liquids as they are a gas because they can be frozen into solids and melted into liquids and vaporised into a gas. In each case the density of the material changes from forming a solid to going into a liquid or becoming a gas. It is about density putting a relevancy between the statuses of materials.

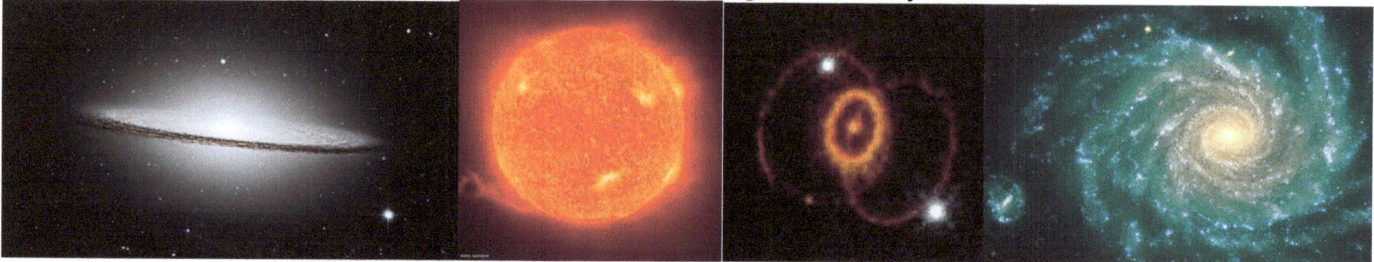

Everything that these pictures tell has one theme and that is heat being displayed in many forms and colours. It is a density variation and those results from objects holding space that moves at different rates in relation to space surrounding objects. Every picture of a Supernova tells a story of the loss of density in the liquid part of the star because materials can only lose density when it bursts by a nuclear explosion. In every cosmic picture we see heat flowing from a less dense and bigger space to a denser and reduced space and that is what gravity is. It is cooling the Universe by applying movement. It is all about moving space to preventing material from overheating where movement brings about density relevancies.

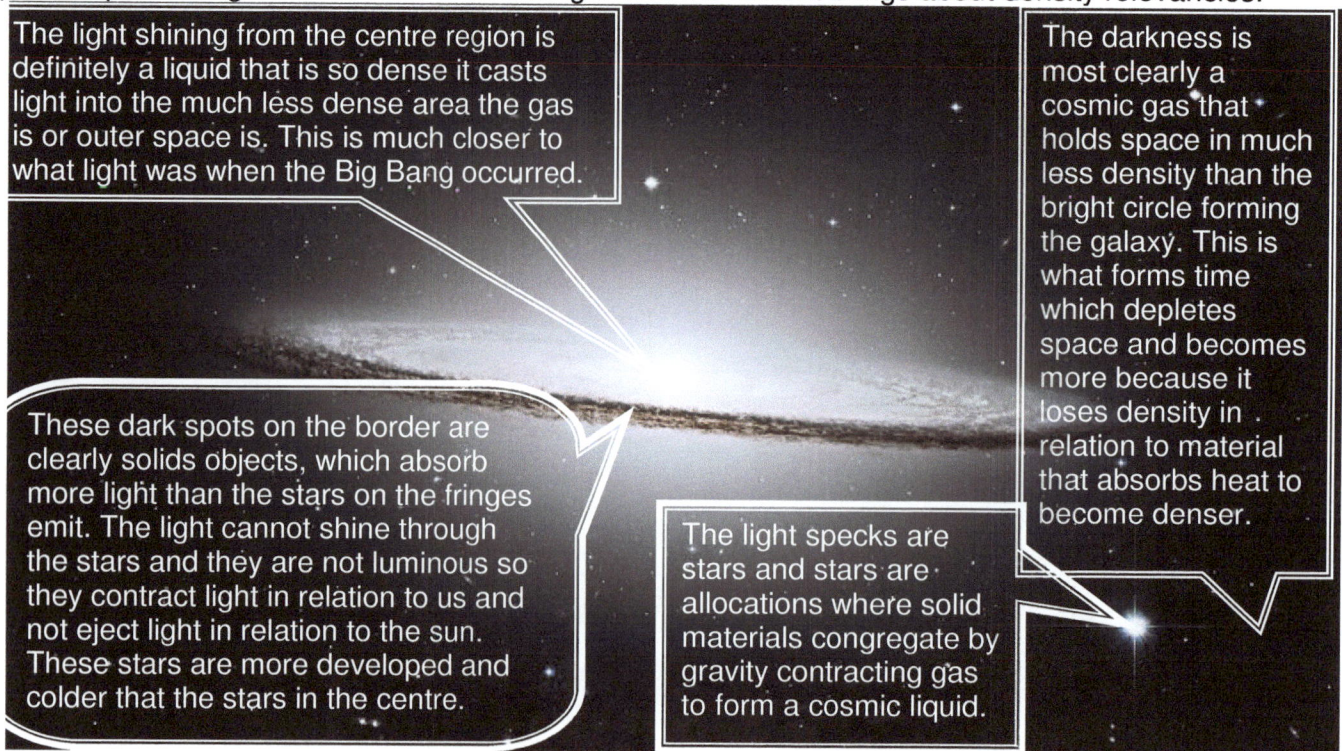

The Universe is about heat finding concentration through movement applying by varying density levels. Whatever is in the Universe is in motion too. If it does not move by moving towards then it moves by moving away but seen from any point every point forming the Universe is on the move. That movement is gravity and albeit moving towards or moving away, it still is gravity. The movement can come from material moving faster than the speed of light or outer space expanding as time moves on and time forms space and therefore space expands in accordance with time growing which is what expanding in the Universe is and this expanding is movement. There was a time before the Big Bang, there was the Big Bang and there was a time after the Big Bang. Before the Big Bang everything there was, was in singularity as it is at present also. Then the Big Bang came and what was then overheated and exploded. But then everything that is was already inside the Universe and nothing but time came afterwards.

There is a problem I see with the Big Bang concept! At instant 10^{-43} second the temperature was 10^{32}K. At instant 10^{-35} second the temperature was 10^{27}K and then at 10^{-6} seconds the temperature was **10^{13}**K. With all this being true where was it zero degrees K. If there was 10^{32} K and 10^{27}K and **10^{13}**K where was it 0K? To have these phenomenal temperatures it had to be 0K somewhere because if not then 0K was equal to any of these temperatures at the time of that specific event. That concept we then carry to the time factor. If it was 10^{-43} and it was 10^{-35} and it was 10^{-6} then there had to be a place where it was 1 hour and 1 day and one year. You can't have 10^{-6} without at the very same instant have another place where the time would be 60 seconds. When aircraft refuel in flight it happens at high speed but in relation to each other the aircraft stands still. To travel at four 400 k / h somewhere then something else had to stand still in order for the aircraft to fly at four hu400 k / h or else everything stood still according to the aircraft. For the pilots flying the refuelling aircraft the one was flying the speed of the other without thinking of the earth passing by at 400 k/h. everything is in relevancy. To fly at similar speeds the speed difference, which is the same, had to be the relative speed is 1 or equal.

That means there is no hot or cold in the Universe just because if it is 42° C on my farm it must be –20° C in New York. I can't have 42°C on the farm and that is it. On the farm it might be 42° C in the shade but believe me in the sun it is 65°C because there are differences. If it was 10^{-43} according to the time applying to the earth then where was the earth? Did the earth rotate 1/24th time around its axis by the time the first hour come to be? Did the earth rotate once when the first twenty-four hours came in place? By the time the first year came about I guess this was when the earth rotated around the sun for the very first time otherwise it could not have been one year or minute or second or a part of a second. This is the incoherent rubbish one gets when using mathematics with a scale applying to science putting the earth in the centre of the Universe. One can't be on Jupiter and apply the spin of the earth to form a concept of time because the time on earth is in relation to the degrees the earth turns in a specific period of time. This humanising the cosmos shows the narcissistic backwardness in the argument formed by using mathematics to draw conclusions.

When temperature was 100 billion Kelvin at Time ~ 1/100 second what then at that time period was thought to be cold or hot. If this is stated in terms of today's zero Kelvin which is the scale they use and that was outer space then this statement is on the verge of insanity because then the temperature of 0 Kelvin was 100 billion Kelvin at Time ~ 1/100 seconds. That means using that scale there had to be regions hotter and colder than other regions were. It is as if they put the Universe functioning in regard of life and life into the thick of things and because we are alive at 37 ° C and that is when outer space is 0 K. That was the Universe then and this is the Universe now and by putting it to an inclusive scale is indicating the universe is comparable. That is insane and can only come as a result of mathematical formulating that does not reason or put values mindlessly together. The cosmos was 100 billion Kelvin and that means there was no 1° K anywhere, which means there never was 100 billion Kelvin. What then was the freezing point of water at that time? Or then answer this: If space was 100 billion degrees at what temperature did water freeze and if there was no water to freeze that Universe couldn't be 100 billion degrees K because then water did not freeze at 0° C. That shows time stood still. Then at Time ~ 1/10 second the temperature was10 billion Kelvin which is a little over 10 billion times more than now as the Universe continues to expand. If this means water was 10 billion Kelvin then there was no water and mentioning this proves madness because there is a relevancy that could not have existed at that point. This goes on and on but this says little to nothing because at that point when it was 10 billion degrees in the shade at noon in the summer without a cloud or wind blowing at what temperature did water turn into vapour to bring about rain as to relieve the Universe from this overbearing drought. This is what you get when you design your personal Universe in accordance with the needs you see befitting a Universe. The same applies for time. Time is the rotation of the earth in relation to the space the earth moves through. How big was the earth at that time to give the influence in seconds that they measure by? With formulas they personalise the Universe by putting them on earth as the applying measure of everything

The Universe works by relevance and to fathom what is big we have to refer to what is small in relation to what is big. In the Universe there is no big as there is no small. The biggest star we know of is a Black Hole and with the density in a Black Hole all the material within the normal Galactica will fit into the size of one atom. In that case an atom is bigger than an entire galactica. Whatever size the Universe was at a time it had a temperature of 10 billion Kelvin, it still is that very same temperature and size that is present today because whatever was that was filling the Universe at that point is still in the Universe and all of that will be in the Universe for as long as we can think. This shows in cosmology there is no hot and there is no cold. There is no far and there is no near. There is no young and there is no old. The youngest star is just as long been in the Universe as the oldest star is because "mass" does not pull "mass" and no star formed a unit by "mass". All material formed when dots formed and space was not yet even a thought. These are man's misconceptions because man makes science revolve around what conditions man would prefer. Man is no factor in cosmology and deserves no place in thinking in terms of the Universe.

This picture does not show singularity or a Universe gone "flat". This picture does not show singularity in any state or form because it makes space inside a picture and singularity is without space because the Universe goes "flat". However, it shows the shortcomings of the formulation of the equating of the Universe by applying breathtaking mathematical expressions. Where is the border that Universe has that we see in the picture? Where is the flat side where the flat square the Universe forms end. Having a top there has to be a bottom and that is space. There is no limit to the Universe but the incorrect mathematics puts a limit on the viability of the formula and it is this limit they transfer onto the Universe. If this is absolute brainpower I thank my Creator for making me as incomprehensibly stupid as I am!

I show do where to find space less singularity The Universe started when singularity heated from the eternal zero it was to the infinite heat it developed. Before heat came there was no cold as there was no hot. That means in the cosmos there still is no heat in hot or cold, only margins forming borders. In nature space expands only when heat increase. Even the Universe at first was not strong enough to contain the heat that rose. Why did the heat expand? This process has a name. It is called an explosion.

It is when heat becomes contracted in the process of developing new space where such containing will light up to produce photons and in the process turn cosmic liquid to cosmic gas. Think of what thermal nuclear reaction came about when far more than half the Universe formed light in heat. It cracked and split the cosmos in two. The other part was contained by gravity to counter act the expanding but much more development came to be light with the heat forming the first photons. Only by heat creation can space grow. There is no other process that can form space other than heat overheating into space that expands into more space to cool down. The initial heat must have been infinite to bring a parting between what became cold and what became hot placing material in between what is hot and what is cold or that which can never start and that which cannot end. To humanise this affair Mainstream science accepts the concept only when it comes to applying their human values of a very manageable and understandable 10^{34} K. Science has "mass" to be clever as to play God to charge the admiration of all other little people. They calculate so that they can admire the mathematical ability of one another as to miss their stupidity of not understanding all the reality that is out there and which they have to miss by looking at mathematics. Their mathematics doesn't promote the Universe. It reduces the Universe to the size of their mathematics.

By moving material removes heat from liquid to solids by leaving behind a cosmic gas. Look at what the inside of stars held before the blowout. Stars are filled with heat on the inside and what the star released with the blowout is heat. It is heat that comes out and that is what a star is. It is an unsealed container of heat and gravity is freezing space into a condensed heat. The reason why we think space is hotter within the earth is because we experience the heat the space releases form the confined space within the earth. But by releasing the heat we experience the space is colder because it is finding relief from the heat it cannot carry. It cannot carry the heat because it is cold. Outer space can absorb as much heat what is cast into it because it is so hot it can absorb whatever overflow comes its way. The cosmos is a variety of density caused by differences in movement within any specific confinement of space in relation to any and all other space formed by heat. I show what forms the Universe is small spots forming dots and these dots when compressed as it was during the Big Bang form heat in abundance but when it expands it relieves heat by expanding into abundant space. That is what the Big Bang was about. Space overheated and formed a means too cool and that became space.

The Universe is formed by heat that we call light and light is dark when it moves away from us (expanding or becoming bigger thus moving further apart and by that it is drawing visible light inwards) and light is bright when it moves towards us (concentrating by gravitational contraction because contraction makes the light denser) but everything in the Universe is a form of heat.

I am about to show that in the Universe there are two substances filling the Universe. One is material that forms clusters in atoms or stars and the other is non-material and there is no such a thing as vacuum or nothing. The Entirety of the Universe is about the relevance of density variation. The faster any object moves the slower will the solid move becoming intensely more dense until it stands still in relevancy to the liquid space that increases in movement where the density will rise to appoint the liquid is as dense as any solid is and it then reaches the speed of light. However at that point the solid will go into singularity and singularity will go into time that is immovable. That then becomes a Black Hole where the solid is immovable and the density removes all forms of what space is in the cosmos. It is the ratio between occupied space and unoccupied space that increases.

There is no mass in the Universe but it all comes down to specific density. Because the faster moving stars associate with more space per time unit the overall size of the star decreases because the space the star runs through increases. However, this allows the density of the star to increase because the material then becomes more solid as it moves through

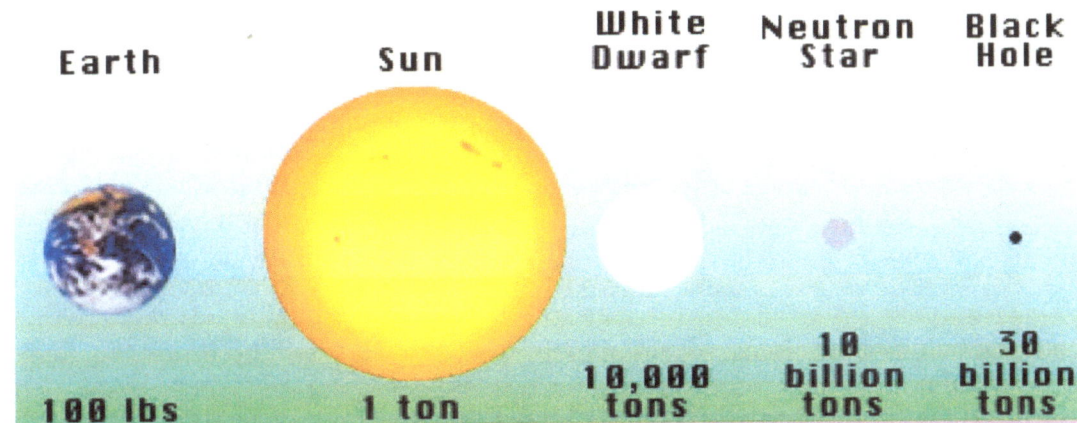

more space. It becomes more massive because it becomes denser but because it becomes denser it becomes smaller in relevance. As the object moves faster it also moves through more space and moving through more space makes the space that material hold become less in comparison to the space that the object goes through while moving that fast. As the space the object goes through increases the temperature of the object decreases and the object cools down because of the movement that increases the cooling of material. It is the same as putting a fan on that blows over the radiator of a car whereby the blowing of air increases the space the water circulates through and the water cools more because the increase in the volume of space moving in relation to the water cools the water more. When any object moves it duplicate the allocated position more as the movement increases. It fills more space in duration of time but during the instant in time it holds less space per single time unit. In this we have space-time because time going singular produces space. The shorter the period of time is that an object fills space the less space it would fill in the instant but the more space it will fill in the total duration. In other words it shrinks the space it is in by extending the space it holds relevance too. The space material moves through increases as the space material holds shrinks. This will go on until only singularity is left. This means the size of the atom inside a star reduces as the star movement in spin increases and this is why stars that move faster or is gravitationally stronger will reduce in size but increase in density.

As any one can see a star is a container of pure heat but it is not a container that holds pressure. Pressure is associated with space and not with heat. When the star or heat container gets pressure it explodes as we all see in Supernova outbursts. The moment the star gets pressure the space it holds increases and that indicates a sudden increase in temperature. If the star explodes by increasing the temperature then gravity is all about reducing temperature because as temperature reduces the space will reduce accordingly. Temperature is not some scale on some meter some scientist devised in human terms but heat expands when it gets hotter and heat contracts when it gets colder. The space the earth holds might seem hotter to us but that is because it is more contracted. To be contracted (not pressured because it is not a metal container such as a boiler is) the heat has to reduce and increase shows heat levels rising. If space expands, the heat increases and if space reduces the heat decreases and that is cosmic science. The levels of heat drops because the heat transmits away from the space in which it are. We can't transform heat values we have to the cosmos. We must adapt to science laws and not the other way around. True science rules say when things expand it is hot and when things contract it is cold.

A star is a cosmic atom and a star is filled with little pumps called atoms. Every atom in the star pumps heat by spinning faster than the speed of light pumping heat from the outside we call outer space to the inside of stars. This process of contracting and condensing outer space, which is expanded heat, we named gravity. The star spins slower than the speed of light in order to harbour atoms that spin faster than light. Every atom is a pump and the protons condenses space or heat or singularity in conjunction with the neutrons and how this works is it compresses expanded heat to smaller space that condenses heat by confining the heat into a smaller area which we see as a star. The star has the condensing ability or the pumping capacity equal to the combined effort of all the protons and all the neutrons and all the electrons within the container we see as a star and in that capacity we see how gravity works.

To unlock scientific truth we first have to dispose of scientific misconception

In the two pictures we are seeing disposing or releasing heat creates space. We may call it plasma or shock waves or what ever, but in the final analyses it is heat turning to space. Whatever you wish to call that which lies between the particles comes from being a solid, then with adding heat, the solid *"whatever"* becomes liquid and that is the white and orange plasma that we find. That white and orange is heat in a liquid form, just as all flames and smoke is heat in a liquid form. But that liquid does not remain liquid because the governing singularity cannot enforce a commitment ensuring the liquid heat remains liquid. The liquid *"whatever"* you call the heat in fluid form then further overheats turning the heat to space. The space created must be equal to the heat reformed. That is a law of energy where energy equals equality everywhere it is. The only "energy" is to transform heat by expanding or by contraction to movement.

If there is movement it can only come from an exchange of cosmic liquid turning to gas or taking cosmic gas and converting that to liquid. But driving anything or moving whatever comes about as a result of cosmic fluid we call plasma and a many other names but it is condensed heat. Heat differentiation translates to movement and gravity is movement. Gravity applies when an object drops by movement or objects keep on tending to move from its position on the surface of the earth to the centre of the earth. The frustrating of movement by stopping further descending brings about mass and that is measured by weight notwithstanding any other definition to try to correct or hide the misconception science attaché to mass. Mass are a measure and not a factor while gravity is a movement and forms a factor that results in mass.

Galileo proved that all things fall equal and by all things falling equal this idea eliminates mass completely notwithstanding the corruption of a feather falling with a hammer in a vacuum. Yes the feather will fall equal in vacuum because the feather holds a different density to air because it is larger and that changes the relevant density it has to the density the hammer has. But a car falls at the same rate as a person and just as fast as the person carry bag. To stop the falling one has to open a parachute and the parachute alters the density of material in relation to the space the parachute confines and it is the density that changes because whatever is tied to the parachute did not lose mass in the process.

By going into the air as a hot air balloon is heated is pure evidence that gravity is because of density and density is the changing of mixture between solids and non-solids. By increasing the air ratio in a form of heat which is what air is and confining this increase within the parameters of a balloon acting as a container the density of the solid material changes in favour of the air and the increase in air brings about a change in density and not in mass and with an increase brought about in the density the solids and the bag becomes a gas where the lot rises into the air as if it is a gas. That shows clearly that gravity is the cooling of space because when lifting the balloon becomes ant-gravity and lifts up. If gravity is "pulling down" then anti-gravity must be "lifting up". Let us humans first detach culture from facts. Take the argument to iron, which we know well. Iron cannot boil, iron cannot flow or bend and iron cannot brake. Iron is an element like all the other elements we know, not one element can do any of the above, in sharp contrast to human belief. As indicated in this book the limits we should find to guide us we ignore for the reason that we cannot see it. We may not be able to ever see singularity, but with intelligence guiding mankind, we do not have to see everything to believe everything. It is because we could not see religion, but still practised religion that set us apart from the other animals. At the start one would find iron and iron in a "natural state" as we find iron on earth being a human produce on the surface of the earth it will be a solid, suitable for man to handle with bare hands. When such a piece of iron is left in a desert in the midday heat, the human hand cannot handle the iron any longer without aid of covering the skin of the hand. Our perception is that the iron became hot, but that is not the case and our view is a culture contribution and not scientific fact. The sun mixed cosmic liquid known to us as sunlight, which is heat into the iron that upgrades the heat part in the iron. By heating the iron artificially with combined gasses (acetylene and oxygen or what ever) we now can over heat the iron to a state of flowing like a fluid. Increasing the heat in the iron increases the non-material section in relation to the material section and by heating it to a point of vaporising the heat part lowers the density of iron so much it forms a gas.

Our human culture tells us the iron now is melting. That is a misconception! Lines forms singularity and it is these lines that holds singularity that forms concentrated heat or expanded space as condense or expand heat and singularity expands heat or concentrates heat. After introducing artificially even more heat with more heat releasing gasses we may artificially form a condition where the iron would become a gas. Again it is not the iron that becomes a gas; it is the space the iron finds itself in that became hot enough to become a gas.

The iron particles remain the same; it is the condition surrounding the particles that changes form with overheating. Important to note is the fact that iron in a solid state will surround itself with solid matter in space applying a solid space. By introducing conditions producing ***more overheating*** the space or connecting between the particles become concentrated heat forming a liquid substance! It is not the iron that turned liquid but the wrapper containing the iron that concentrated so much it formed liquid fluid by the introducing of more heat to a point where the overheating created a fluid. It is considered that the oxygen burn and by that the iron heats up. NOT TRUE! If oxygen burns no oxygen would be left on earth by the time man arrived on earth to use it to the benefit of intelligent life. The oxygen remains oxygen while the oxygen merely does a task in nature where oxygen carries heat to a specific space. On the other hand it is the task of nitrogen removing heat from the point of overheating by means of flames whereby it creates space. One can feel the "wind blowing" as the flames generate created space. In the extreme the creation of such space we call an explosion. In the process where the space between the iron particles still further overheats, it becomes a gas. It cannot be iron that becomes gas, because depending on mixing heat with iron it will be as much a gas as iron will be a liquid or a solid.

It is the space covering the iron particle separating the different iron particles, which will convert and sustain form. The gas is as invisible as space because the gas is the form space holds. It is the relation that materials form not by heat or cold but by linear or circular motion that forms density. There are two forms holding substance earth (solids) created and heaven (heat or gaseous/liquids) created. These are the only two forms of substance that is the Universe solids and non-solids, where non-solids are what increases to form liquids and gas. It is not the solids going liquid but it is more of the liquid in ratio with the solids in between the solids that make a structure go solid or gas. There are cosmic solids and comic non-solids. It is movement going at or slower that the speed of light or solid atoms going faster than light.

Iron is a solid. Introducing more heat the iron becomes more a mixture between liquid heat that reduces the density to the point of concentration where it became a fluid. The iron remained what it is, neither a solid, nor a fluid nor a gas. By introducing more heat it becomes a gas. The gas we cannot see because the gas is space. But so was the fluid space.

The introducing of heat brought about the turning of a solid to a liquid to space and every time more space becomes part of the picture. Iron is in its normal form a solid. That means the space, which the iron particles are in is solid and that disallow the iron to alter the form in which it is. By introducing considerable heat the iron melts changing the form of the iron from solid to liquid. One will find that whatever group one chooses there are gasses and there are solids. If mass was attracting mass then the strongest mass must be attracted to the strongest mass and the least mass must float in the air. $F = G (M \times m) \ r^2$ hardly can even begin to explain the fact that there is a gas that is more massive than iron but floats in the breeze just as hydrogen which is the least massive element.

The entire Universe forms by heat. The Universe started the very instant when cold and hot parted and in between a cosmos came about. There was what was to become dense substance that we now call material or occupied space and then there was unoccupied density that moved slower in relation to the density and moved fast. Today we have three limits with two divisions which forms as solids / liquids and gas. But make no mistake it is all singularity or heat that moves at different rates to establish different densities. That is the Universes.

For instance:

Nitrogen 7	**melts at -210^{0}C**	**boils at –195.8^0 C**
Oxygen 8	**melts at –218.8 ^{0}C**	**boils at -183^0 C**
Fluorine 9	**melts at –219.6^0 C**	**boils at –188.2^0 C**
Neon 10	**melts at –248.59^0 C**	**boils at –246^0 C**
Sodium 11	**melts at 97.85^0 C**	**boils at 892^0 C**
Magnesium12	**melts at 650^0 C**	**boils at 1107^0**
Aluminum13	**melts at 660^0 C**	**boils at 2450^0**
Silicon14	**melts at 1412^0 C**	**boils at 2680^0 C**
Phosphorus 15	**melts at 44.25^0 C**	**boils at 280^0 C**
Sulphur 16	**melts 119^0 C**	**boils at 444.6C**
Chlorine17	**melts at –101**	**boils at –34.7 C**
Argon 18	**melts at –189.4^0 C**	**boils at –185.8^0 C**
Potassium 19	**melts at 63.2^0 C**	**boils at 760^0 C**
Calcium 20	**melts at 838^0 C**	**boils at 1440^0 C**

Every element clearly has a different density point It is not mass that plays any role in the measured formation of elements. Elements can be on earth prone to form a solid better than it holds a liquid form but that is related to gravitational conditions we have on earth. If man would land on Jupiter there would be no man landing on Jupiter because the conditions on Jupiter could never sustain any form of life. All the above elements would have total different for limitation on Jupiter than they would have on Earth. If one looks at mass there is no connecting point putting a realistic ratio between what is heavy and what is light? If gas floats in the air we have to presume it is lighter than air because if something floats on water the density the object has to float makes it less dense than water is and therefore it is lighter than water is. In this same manner we have to look at gas and irrespective of the "mass" value Newtonians grant the element to have, if it floats and it is airborne like a gas is then it is less dense than air which makes the element lighter than air. The hot air balloon fills with more hot air and that reduces the density that allows the balloon to float in the air. The balloon is lighter than air notwithstanding it never lost mass.

Looking at stars the way Newtonians do they relive the coal burning boilers. They see coal furnaces being stoked to burn and heat boilers. In the days of Newton coal stoves were the nuclear science of the day and while all other departments in science moved on and away by developing away from 17th century values and from coal stove principles Astrophysics and cosmology remained true to Newton by reinventing the coal stove in so many ways not even the coal stove could think of the facets it can go through. Newtonians see stars being fuelled like coal boilers and such steam boilers can run out of fuel. This is so much Newtonian backwardness as mass forming gravity and the moon coming closer and the cosmos shrinking and we falling into the sun because of non-existing dark matter making up what is required to make Newton not to seem the idiot that Newtonians are because they make him and his contraction theory to be less foolish that what it apparently is and they overbearingly are. What is of vital scientific importance is that there are three fundamental dimensions controlling the universe. The three are beyond intermingling and one confirms a status in relation to the others but not intermingling in status. From singularity comes matter and forming space-time in own accord. By matter not controlling time, space grew uncontrolled and the third dimension came about. That dimension birth we now recognize as the Big Bang, but the Big Bang is the last of a three prong cosmic growth. Science has to recognise the dimensions of densified (singularity), occupied (matter behind the electron) and unoccupied (space-time outside the orbiting electron boundaries) forming three points of cosmic recognising space-time.

Every atom holds (I am guessing), as many dots as the sun has subatomic particles per atoms and that would still be a very conservative guess. Every dot is a controlling centre selecting a regional centre where every regional centre selects a centre. This goes on as long as there are spots forming groups as individuals unable to survive independent. The others that was unable to group formed heat that became space, which became the broken dots. The dots form groups to survive and as a group, the survival depends on doing what the group has to do to remain cool. In another book, I reserve one chapter to explain the phenomenon what I called the Lagrangian atom. These dots arrange in a manner that they could favour either the space duplicating aspect or the space dismissing aspect.

This can only be the result of the fact that even in the case of the sun, the inner space is almost entirely liquid heat and the liquid heat produces sufficient space to dismiss as the centre that holds the heavy metal particles, where all the dismissing is done. The liquidity provides motion while the solidity removes motion in the centre of the star. The dismissing going on is in the space factor where the space leads to a denser heat within that space because there are insufficient material to accommodate all the heat by the dismissing factor T^2. In that case motion far outweighs dismissing $k>T^2$ but a time comes in every star that the dismissing takes absolute charge. $k<T^2$ That is when the star goes dark. The Earth is mainly about duplication of space much more than dismissing of space and so is every structure in the solar system.

I would suggest we think of stars in the following terms. A star that generates and transmits a lot of light is weak on gravity because their progress started recently. They command a lot of space-time but the demand they have to keep their cooling acceptable is very low. In that they can generate a lot of light but with the demand on cooling low and the gravity in the centre not very developed, those stars cast a lot of light back into outer space. It is just because of the size the stars holds that tell that the stars are still young and have a weak developed governing singularity. The stars will have very prominent hydrogen and helium layers, with the inner core not very prominent. The control of the star is still very much in the individual atoms and in that the motion the atoms have to produce in order to maintain their individual singularity will only come about through motion. The atom has to make contact with as much space-time through motion as possible since it has a very poor ability in contracting space –time in support of the cooling system. There cannot be something big or small except in the relevancies of perceptions and then the relativity of such perceptions becomes questionable. There cannot be hot as much as there cannot be cold The sun freezes hydrogen to a liquid at 6500 ^{0}C and outer space boils over at 0 K. If we humans cannot or will not abandon our human culture driven perceptions and our mankind's pre-programmed perspective we may as well return to astrology for what the future hols. There are so many boundaries out there ready to destroy us because of our lack of insight, as did the challenger disaster. Creation birth started off with one dot so small eternity met infinity within. Then came one more, and another and they continued coming until there were a countless number of dots. The accumulative size of the dots were the same size as one dot because in the true Universe big and small plays no part. The dots were infinitely small and eternally big at the same time because size is a relevancy and without one the other has no size. So in the true perception, there is no difference in size.

The idea than humans make aircraft that can fly has no meaning in the cosmos. Life as an entity is totally alien to the cosmos and life can only be on earth and nowhere else unless the atheists can prove otherwise. Everything that moves in the cosmos moves because of gravity and gravity is a difference between densities in space. A gas is a lesser density because it spins at a higher velocity and not because it holds more or less mass. A gas floats in space because like the hot air balloon it has a higher relevancy to heat or it captured more heat in-between the atoms forming the compound of the element. When anything moves fast it is colder and because it is colder it captures more heat to maintain the speed of spinning it has. On the other hand when objects spin slower the elements tend to be more solid and the density is much higher. Every element is made up in form by innumerable many dots that form singularity. Every dot was by itself as well as the accumulation as it currently is the present universe. The earth in itself is a Universe standing apart from other universes such as the moon as well as the space between the moon and the earth. The moon is a universe. Rules applying on earth do not apply on the moon and visa versa. When considering conditions with in the oceans and applying space-time another set of rules apply therefore the sea places a body in another universe. It takes the same engendering technology going underwater in deep sea diving that going into outer space. Every dot was a Universe in its own and the accumulation was a Universe. The earth in itself is a Universe as the moon is a universe, because rules applying on earth do not apply on the moon and visa versa. When considering the conditions with in the ocean and applying space-time another set of rules apply, therefore being in the sea places a body in another universe. The number of universal entities is still countless, as much as it was in the beginning, before dots formed atoms. Every dot insignificantly small as it may be, is a part of another Universe as much as it is part of the accumulative Universe and every dot in the infinity holds singularity, which we translate as " nothing" being " darkness".

The very first instant when the cosmos started the perfect became imperfect. When what was perfect became imperfect the Universe moved as time in eternity split from time in infinity and I show where time in infinity is and where time in eternity is. When the spot differentiated and became differently allocated from the dot the Universe started. When infinity moved away from eternity the Universe started. When the perfect overheated, hot and cold formed relevancies that put space in between time in infinity and eternity. Even today this is the fuel that drives the Universe as liquid that parted from solid reunites with solid to form a density difference. Every dot insignificantly as it may be is a part of another Universe as much as it is part of the accumulative Universe and every dot in infinity holds singularity, which we translate as "nothing" but it cannot be nothing. Singularity is what forms the Universe and is the smallest that can be.

The light specks we see scattered throughout space at night are stars and stars are allocations where solid materials spinning and with that are by gravity or movement contracts gas to form a cosmic liquid.

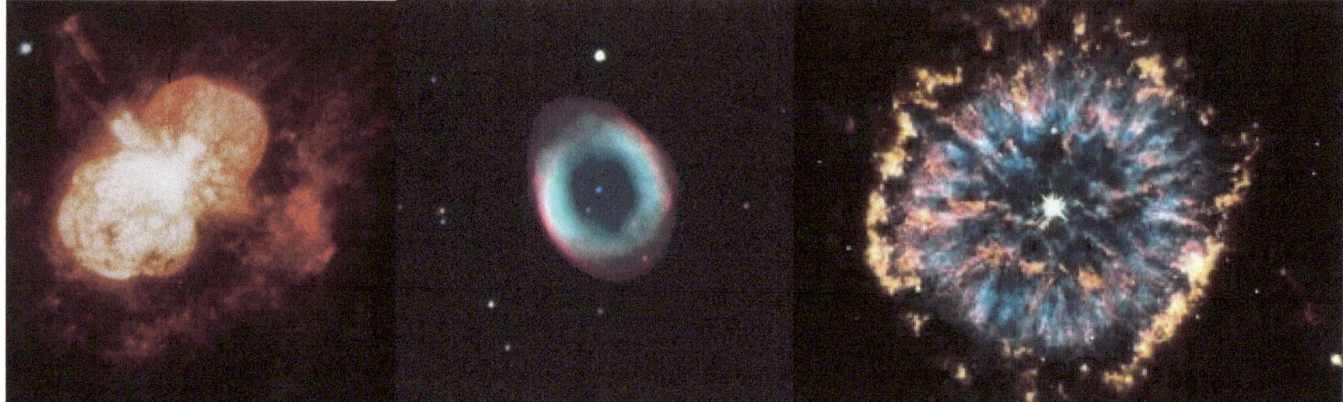

It is clear from all the images that the liquid that was inside the star before the explosion froze in liquid

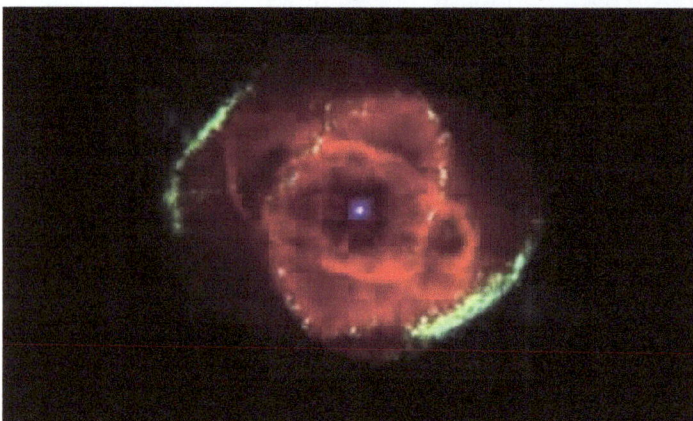

form when the explosion thrust out what was inside the star and clearly it was heat forming space in the form of liquid that came about because what was remained on the rim of the circles. The liquid that was inside clearly remained liquids that froze in space resulting in liquid forming the outer edges of the layers and as it turned to gas the gas formed more outer space. Clearly the density variation is visible as what was inside formed either a denser liquid or a darker gas. It is also clear that one layer after another layer overheated and expanded by causing an explosion as we can see from what remained.

This explosions and what comes about from it as "shock waves" carry many other Newtonian names but in the end it is heat going into gas as the cosmic liquid forms cosmic gas and cosmic gas becomes more cosmic space as it expands. In the picture of the sun the sun shows liquid and so does the galaxies and it is all too obvious that space in the form of gas freezes into liquid as gravity reduces the space into frozen liquid. Moreover still, we can see the point of the remaining core holding the incredibly dense material that forms the singularity within the star. Every star is a Black Hole in development. As the Universe grows the density of material increases by the same margin as outer space loses density and in the end the core of the star unites all atoms within the star where the atoms by movement becomes singular in one unit. Inside every star spinning there has to be an iron core to produce spin because I prove mathematically that electricity is the very same as gravity but it is dimensionally charged on

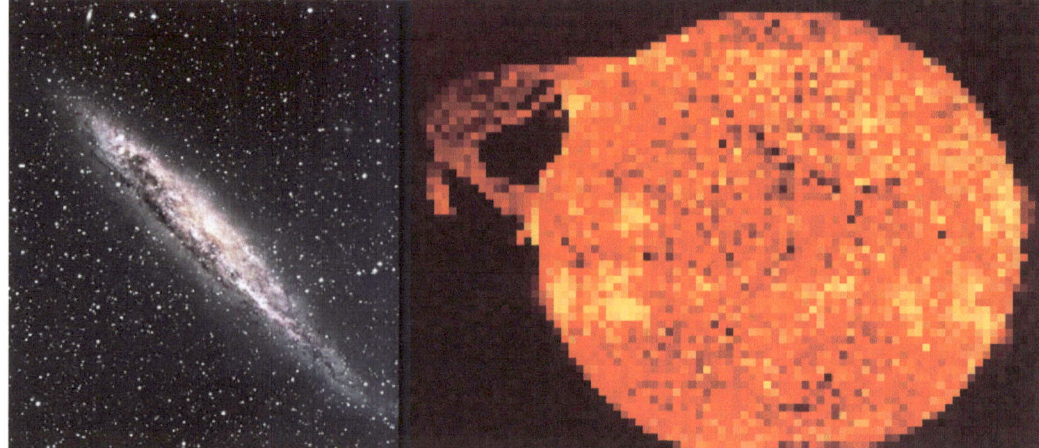

different levels. In every pigment of every picture we see heat being in contrast to heat setting differentiating levels in any volume of space one compare to another. The entirety is heat being space and heat being condensed but in the end it is heat forming a different contrast to other heat that forms more contrasts. Every atom forms a Universe. The sun concentrates heat towards its centre and the galactica concentrates heat within its centre and then everyone believes mass pulls mass without ever realising what drives the Universe is heat contracting and heat expanding. The driving force if I may use such a primitive word is relative density. The star collects heat by turning around within heat and that turning makes the star denser than outer space because by being denser the star is also colder. The star grows by collecting heat because the star is a unit of dense material and outer space is a nonbonding substance of non-materials. Material grows while space expands and it is about density relevancies and the density applying varies between elements, atoms and gravity. As time moves on it forms space and that process is the fuel the Universe applies to return the Universe back to singularity as it was.

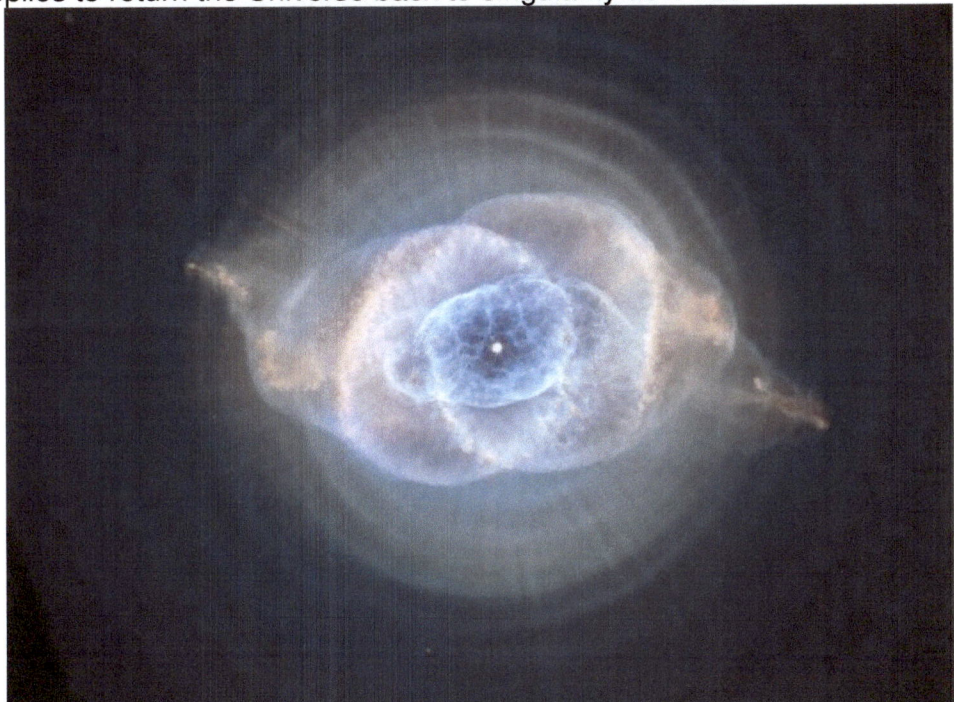

Look at the image of this Nebula and see how it glows with heat. It is clearly show the star was a heat container that opened up and had the liquid inside burst into gas. My eyes tell me this picture is a showpiece of heat contained and the container overheated where the liquids turned to gas. In order to cool gas the gas has to flow and our refrigeration and air-conditioning work on this principle. Even the engines of the space shuttle apply this very principle to cool the heat. The star is hydrogen gas frozen to liquid by the spin of the star and in that the star is filled with liquid because the pumping of the star froze the gas into liquid. By contracting the heat surrounding the star the heat condenses and it releases heat because it became cold but due to the spinning slowing down it couldn't maintain the cold and the supernova overheated.

At first when I started a study to find out more about "gravity" and the ever, elusive graviton, I came across lazy gravitons, and eager gravitons. It seemed that the lazy gravitons produced little gravity and the heavy elements such as Xenon and Radon are the heaviest inert gasses and therefore must be the laziest gravitons around while Lithium as a solid being the lightest solid must have gravitons so potent you can use them as Superheroes that replaces Tarzan on weekends and Public Holidays. This remark is pure trash only because the graviton is pure trash.

Gravity is a reality because atoms pump heat into material as to avoid overheating that result in an explosion as we can see when looking at Nebula. There are so many facts that accepted science know about, but do not fit into the perspective of accepted science, after which science blatantly ignored these facts. I have indicated but a few examples and the examples are the most basic science offers. All information I have disproved thus far is science taught to children at school. Therefore, it is not hard to imagine how much nonsense they propagated in the more complex issues. Can you imagine for one second, a star that "collapse under its own "gravity!" ...And this rubbish resulted in that two Nobel Prizes were awarded to two Nobel winners because they proved it!

Only in Newtonians' Science as primitive and backwards the outlook is will a coal-stove fire need a thunder- storm to start the fire. Please allow me to explain, and believe me, I am not making this up as I go along, just to ad humour to this otherwise very dry letter. I did not invent this fiction in order to create a bit of comedy; this is truly Xepted Science.

There apparently is a point at which point the star performing like a coal stove starts because gravity lit a fire and the fire is burning the coal (core) to get the star going. Take into consideration that every idea science clings onto today was an idea put forward when ships were driven by winds blowing and in Africa there were believed to be dragons.

This is pure hot science as was practised in the 17th century but very current. At the time everyone saw steam changing as advanced as nuclear science is viewed today. They new the boiler and steam was going to change life and the essence of it but that was then projected this massive new knowledge of the day into astronomy because only the brightest minds could understand the working of steam. If it formed the edge of science astronomy had to use it. So steam was the fuel of the Universe because Britain was the centre of the Universe. As in the case of all fires, the star (or stove) ignites with a small spark. It had to be that the fire grows as the hydrogen heats up, just as it would do with the coal that has to heat before it can burn.

Depending on the amount of fuel available, and the heat to ravage the fire, this star can burn from anything between a few million years and eternity. What a lot of Neanderthal bollocks fit for fairy tales and other fantasies. The fuel the Universe holds can never run out because it is time forming space. They thought it is the mass that determines the life cycle of the stars and the bigger the star the more mass it had so the shorter would the lifespan of the star be. Today we know big stars are very soft in terms of mass while the smaller the star is the more potent it would be in terms of mass. This is because the relative density of the star increases as the overall size of the star reduces. This is all determined by the "weight" of the matter in the star. In the case of the sun, the star will "die" leaving behind the ashes of a helium core, with the density of 10^{17} g/ cc. How helium can remain as helium with such density, is yet another unsolved riddle. Every idea Newtonians believe in is centuries old. Whenever they discover something new such as the Titus Bode law, the Roche limit, the Lagrangian points, the Coanda effect, the Hubble expanding or nuclear physics it is burdened by either covering the importance, hiding it by preventing it from becoming common knowledge or turn it into a prank. Hot air lifts balloons and when this clashed with Newtonian mass pulling things down it became a joke by referring anything that overshadow its purpose to be hot air. Can you believe that they can believe a star can die? Only life and what comes from life can die because life holds time at a limit but not the Universe. To their view the Universe is limited but the cosmos shows it is eternal. Black Holes prove that stars can't run out of fuel!

So ignoring hot air they stuck to the story of mass pulling things down instead of thinking that if hot air makes things rise it must be cold air that makes things drop and that would have brought my theory to the forefront two centuries ago! To put this into context, they had the Newtonian stove burning coal and the stove starts with a small match, then it simmers, while the heat produces more weight and pressure and the coal burns until much coal burns bright in all splendour and glory.

With the fire raging, the flames will light up the night sky, and the light is visible for miles around. The more the coal the stove has the more the glow is of this red, hot fire. After a while the stove runs out of coal, and with no one to stoke it, it starts to simmer, after which the glow disappears altogether and the stove dies out. Believe it or not but this in all the out datedness it represents, this is still modern cosmology, preached by people that has so many degrease they can put it to use as wallpaper to cover walls!. It is clear that the twenty first century has not reached cosmology, because if ever I heard a Medieval, old wives tale, then this must be it.

However, this is not where the stove comparison ends. I have books in my possession, where one of the worlds most accomplished and renowned cosmologists, is of the opinion that the hydrogen falls on the red, hot core of the star and then ignites. This will happen to cooking oil that drops on the red, hot stove, but not on the inside of a star. Science only chases money! Somehow, somewhere someone found traces of minerals on the red planet, and this unleashed the feverish interest that is taking place currently. For many decades and years, there was a lack of interest in manned space flight albeit to the moon or Mars. The only interest was getting into space and manufacture semi conductors in "micro gravity" again to profit the Hoggenheimers and the John Dows of the world fit the bill with Tax money.

I might sound cynical, which I am not. On the contrary, I applaud every move NASA makes all the way. All I am asking is for honesty. A small amount of earnest will go a long way. In fact, to my thinking, they can take 100 percent of all the money spent on arms each year and dedicate it to NASA. That should not be that hard to do. Tax all sales of arms by one hundred percent, no matter who is baying be it the American, Russian or Chinese government, or the African governments living on charity, yet having enough money to conduct the most horrifying wars. Less people will die and mankind will profit much greater from such an action. Whatever we see in pictures about the Universe we see density differences everywhere.

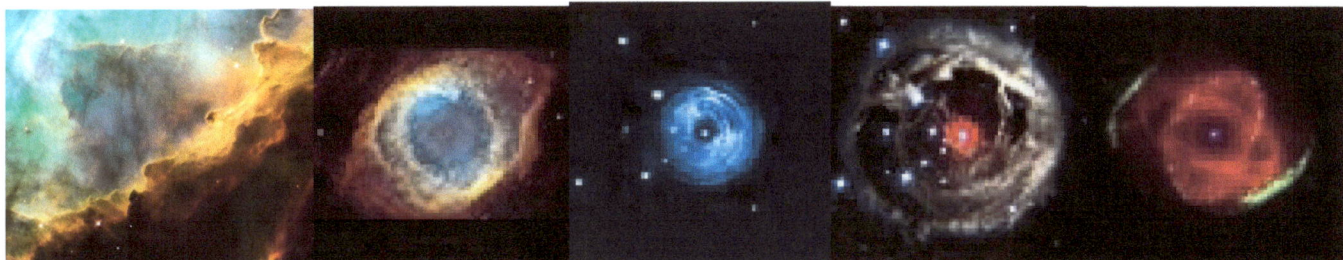

The density by gravity shifts from where the substance is nonbonding to where the heat as a substance is bonded by movement forming materials. The density shifts from non-materials to materials and this flow of heat forms one part of gravity. As the liquid or cosmic gas loses substance to materials contraction the density in the non-material division loses compactness by losing compound and therefore it expands into space while materials collect heat and therefore grows in space. That is what the Hubble shift is about. As materials form density the non-materials loses density and also therefore value. A star does not grow into a Black Hole but outer space envelopes the star into a Black Hole because as the star seemingly shrinks so it is outer space than reduces the effective space that it holds. The more outer space loses compound and gain space by losing density the denser does material get but as it gains in density it loses in relative space. The cosmos is a shift in density running from non-materials to materials and this evidence is proven by the fact that the earth grows in size while the moon drifts further away and this is correlated.

In the star heat serves to keep the atoms cool. Without sufficient heat flow the atoms will get hot and expand. When the atom expands it not only holds more space depreciating the relevancy of space to material within the star but it spins less and the heat within the tar rises even further. This forms more space as heat that expands by overheating produces space and when the space reaches an ultimate critical level the Roche limit is reached and following the Roche limit come the Roche lobe. I explain this process in much better detail elsewhere in this book. Therefore the fuel that a star runs on is not coal producing fire that can finish when the stoker retires from his job with old age and it leaves the star to *"die"*. No star can ever *"die"* because a star is not a coal stove burning as it cooks oil or burning hydrogen. The atoms are pumps pumping liquid into the star and the liquid allow the material to grow while the liquid produces the rise of relevant density where the relevant density matches the overall density applying in the Universe at that moment. A current Black Hole was a star a very long time ago and our sun is a future Black Hole a long while from now but it is in the process of developing.

Looking at the stars and galactica through a Newtonian perspective Creation rumbles on without perspective, purpose or destination. It is a tragedy that people will be so obstinate in their programmed mentality to click on one thing and miss the entire picture. Only a definite relation between two balancing values forms the complete Universal relevancy of SPACE-TIME $a^3 / T^2 = k$ as Kepler got it from the Cosmos. It's clear that the centre concentrates heat.

I compiled **a new cosmic concept** by which I eliminated all the incorrectness that Newton has burdened science with but with this being my opinion I did not find a garage full of academics supporters waiting to applaud me and to uphold my views on the matter. Gravity rests on movement of material in relation to other material also moving. The movement of the sun provides the earth with movement but not only that al movement going straight becomes circular movement and circular movement takes place within the circle in which it moves going forward as a straight line and in that idea of a circle becoming a straight line and a straight line becoming a circle the entire concept of cosmic gravity is vested. According to the Big Bang theory the Universe expands and there is no evidence of pulling bringing about a Universe contracting or becoming smaller. The Hubble constant is sole evidence of this proof of expanding Therefore I challenge the concept they build on the fact that mass attracts mass and everything is pulling everything else. Yet still I was not going to be ambushed by their relentless stonewalling my efforts and blocking my efforts in introducing both the incorrectness and the new cosmic theorem I concluded. My views are founded on what is there and what I can calculate as it is applying.

Their mannerism in blocking and frustrating my opinion when showing the mistakes in science convinced me about a Conspiracy in Science in Progress and this spurred me on to tell the entire world about their brainwashing students minds. By the manner they selectively withhold information when teaching science, amounts to deliberate brainwashing of students in physics by "normal" education practises. The new concept I wish to introduce puts all emphasis on space ands material is only space filled with material substance while other space is filed with non-material. In the end all space are equal but the movement it has makes the difference it presents in relevancy.

All space structures hold in the centre singularity concentrating heat and from that centre that all material holds comes all the drive. I can go on and on but heat in the centre couples gravity to space-time, just like Kepler said before he was spoken for on his behalf and without his permission or his agreeing to it. It would have been much more palatable if the Newtonian views were based on some form of a possibility but it works on total fabrication of facts that has one purpose and that is to mislead and to mesmerise by concocting untruths in the use of untrustworthy mathematics and meaningless formulas that no one can legitimise or prove! Newtonian physicist know all their mathematics are senseless because it proves only that they understand nothing except it makes them feel equal to God Almighty because of a feeling of total superiority the mathematics allow them too experience. This we can see from how they tell the Universe what they want it to be. The flowing is the result of using much superior mathematics to prove they know nothing about what they know.

If I made a statement that Newton is wrong about gravity, which person would believe me? If I said all those in science know very well that Newton is wrong about gravity but is hiding this fact for personal benefits in order to ensure their work remains to be accepted, who would believe me. If I said that everyone in science are aware that the formula on which all science are based $F = G\dfrac{M_1 M_2}{r^2}$ is as false as a politicians' honour, this fact will then come as an astonishing surprise to everyone and I get blamed for smearing the characters of the most honourable group of persons God ever thought to put on Earth. .

This $F = G\dfrac{M_1 M_2}{r^2}$ is the formula judged to form the basis on which the entirety of physics rests. Yet, nothing can be more inaccurate than this formula that science forms its entire basis on. As students learn this formula off by heart it starts off a brainwashing process and this procedure becomes everything that science represents. Should students not accept this formula as the gospel truth and as if it forms the only concept that could represent accuracy found in the entire Universe, that student would be sent off branded as not to be capable of understanding the fundamental basis of science. That student will go home labelled as stupid without any further possibility of studying physics in the future. I prove not only that this formula is rubbish but also that there is no mass at all.

Every one sharing the Newtonian vision of a contracting Universe is dreaming of a Universe where the lot would one day again come together and Creation will end where they say Creation started some time ago. The presumption is that the Universe has mass that is pulling mass towards one another and we are in the centre of an ever shrinking Universe. That is what this formula $F = G\dfrac{M_1 M_2}{r^2}$ represents. That is what the lot of us must think we know… we think we are forming the centre of the ever contracting cosmos where every Newtonian can vividly see with his or her eyes through any telescope that all Newtonian minded scientists are sharing the centre stage of the ever collapsing Universe. Newtonian science holds the view that the Universe is about to end where all mass contracts into one huge lump of material or that is the basis for Newtonian science.

If you want to go anywhere you better use heat to give you "energy" which "energy'" is another word for heat that will allow you to move. You move with using heat. You need hydrogen but you don't burn hydrogen. You need propane but you don't burn propane or the oxygen that you also need. Hydrogen, oxygen, propane are gasses and because they are gasses they hold much more heat in association than does a solid such as tungsten. You can't burn tungsten as a liquid because tungsten as a sold holds very little association with heat in comparison the mentioned gasses.

It is the heat that the gasses hold captive that releases and that turn to fuel but it is cosmic heat. The gasses as a liquid in association with the solids atoms hold cosmic heat in relevance to a specific density and by releasing the heat the gasses hold in liquid and allowing the liquid to expand to gas it forms space and this release of space is the cosmic driving force used to bring about movement. If you want to produce movement you have to produce heat by converting cosmic liquid into movement that changes the cosmic space in the process. You have to convert cosmic liquid into cosmic gas but first cosmic gas has to convert to cosmic liquid and that is where the sun supplies us on earth with the only fuel we can use. We can use elements that naturally capture the comic liquid and store it or we can use fossil carbon that as life converted the cosmic liquid to reusable fuel back before stupidity ruled the earth but when it releases from the carbon it converts flames (cosmic liquid) into heat (cosmic gas) and it is this converting that is the process we harvest but it still remains cosmic liquid turning to cosmic gas or the other way around. Electricity is converting cosmic gas to cosmic liquid. It is the same as gravity.

In cosmology there is only one method of creating energy and that is by a displacement of heat in terms of cosmic liquid spinning and allowing material to spin (7/10) within a liquid. If the liquid moves the relevancy changes to (10 / 7). It carries four different titles but the application is the same as the four are the same transferring of heat and when Einstein searched for the common denomination between gravity and electricity he was on the right track. He was correct but never found out why because (I guess) he was looking for mass and mass related ideas while that is because every atom is a generator of electricity. The atom has an electron and an electron is what provides the flow of electricity. There is and there can never be any free electrons because there is no such a thing as an electron. An electron is a gaping hole into an atom and it is the spin of the atomic liquid neutron that pumps the cosmic liquid into the atom by reducing the space as it increase the speed of flow from what was below the speed of light to a point where it was an electron flowing at the speed of light and from the increase the velocity of the flow of displaced heat to become higher than the speed of light. Where an electron is we find space-time accelerating from as fast as light to faster as light and yes Einstein is wrong because there is a transfer of heat exceeding the speed of light and going faster than light. The inner atom spins faster than the speed of light and the release of this heat we call nuclear energy. That is the first form of gravity or electricity or electromagnetism, which is nuclear energy. All these mentioned above are the same thing.

If the atoms individually generate a flow of heat we call electricity because when charging electricity with any dynamo it is iron spinning in copper ending the Ferrous process or the conducting of electricity from iron being magnetised to copper and in copper the flow ends where the line that flows is cosmic heat. To explain this in detail by mathematics is too cumbersome and time consuming . When all the atoms in a star spin altogether with unification acting as one cosmic atom (a star) and therefore acting as one unit we call it gravity but it is the concentration displacement uniting in centre- singularity that forms the difference to the measure of the very same applying principle. If it is a flow of electro magnetism it is one electron. If it is the flow of a generating of electron by spinning iron in relation to copper we call it electricity. If it is the flow of all the atoms forming the planet or star joins we call it gravity. What is needed to get the flow of heat conducted is the spin (7/10) of material in space. That is gravity, electricity, magnetism or electromagnetism and by name it sound different but it is the same thing but operates under other dynamics. This flow of heat is a liquid cosmic substance that we call air if it is condensed and outer space if it is cosmic gas and if it is pure liquid we call it lightning, light photons or electricity or just simple flames but it is a flow of heat. When it is condensed by electricity of gravity it condenses from a space valued at outer space that has a 112 proton-displacement to the inner space, which is a 56 proton-displacement, which is the proton value of iron. This limit is set by light that holds a space value of 3^3 in combination to a displacement value of $3\Pi^2$ and these are the two forms of light that forms the Universe.

The light value of $3^3 = 27$ is space that builds the Universe which Newtonians can't see and therefore named it as "nothing". It being "nothing" or not it still it is light moving away form infinity and towards eternity and therefore moving towards the future as it has been doing since time split infinity from eternity fills a Universe to an overflow. The building of the Universe since the Big Bang where the heat release came about as three dimensions formed $112.79 = \dfrac{7\Pi^6}{10(2 \times 3)}$ as space. This is outer space that formed at a value of 112 and inner space displaces into singularity at half that which concludes when light (($3^3 = 27$) +($3\Pi^2 = 29.6$)) combines to form 56.6 and because of that iron $_{55}$ and copper$_{62}$ combines as a displacement of 118 and as the average value ($\Pi \div 2$) which in this case is 118/ 2 and then forms a displacement limit at Cobalt $_{59}$ and the light forming a total value of 56 halves the space-time value of the displacement of 112. I can give you the other formulas but it would be meaningless if I don't explain the entire process mathematically and that is tedious and involves a lot of mathematical principles applying. When a planet or a star does this it is gravity because then the planet or star becomes the value of all the turning of atoms in the star and the flow of heat from cosmic space or outer apace to the centre iron core forms what we see as gravity.

Einstein knew there was a connection but was never able to figure this connection out and he died a frustrated man feeling he was a failure in his own eyes (not in mine but according to his interpretation) and photos taken of him with his hair standing and all messy was from this era. He knew there was a correlation and it formed a direct link but he could never solve the problem and consequently died as a frustrated individual. I was fortunate to formulate the mathematical equation that showed why and how electricity and gravity is the very same thing and to put electricity on par with gravity is the only step in the correct direction. The main issue is material mist spin in relation to each other and this displacement albeit concentrating or expanding is the fuel that drives everything we think of as or we think form the Universe. The process where comic fuel is being heat applied to create movement and that it is the only fuel in the Universe is a fact of life because it even sustains life. Life drives along by using cosmic fluids. Cosmic liquid is what life uses as fuel to function for a short while. Yes, you eat food and you breathe oxygen and that is so Newtonian I choke when hearing this simplistic childish reasoning. It is because of such simple-mindedness that atheist thrive and prosper and feel kings with no one taking on their stupidity. Even the essence of life depends on the capture of sunlight and removing the sunlight as cosmic heat from our surroundings and supplies that to life where life as carbon with life's ability to manipulate what the cosmos applies, life stores this heat in its carbon fibre to be reused by other life forms so that life can sustain life. In short animals eat grass and we eat animals

. In order for human life to live, life has to devour what other life forms left as their heritage and the result of what they did with their time they left behind. Other life leaves their time they lived behind as material with carbon they structured according to what they are while being alive either as plants or as animal meat. The grass grows by accumulating and processing heat from sunlight taking from the air carbon and mixing this recopy to feed animals grazing on the grass. We more intellectual life forms then consume animals that live from grass because with our intellect to sustain we can't consume grass alone. We have not got the time. Time is life…there is just so many heart beats per minute and we have just that many heartbeats that forms a life-time and that translates to time equalling life. Grass won't feed intellectual life and in essence the heat that I am referring to is actually time which is the substance or the result of what time deposited as time moved on. That is why the moon and earth is moving-apart. It is not moving apart it is time leaving space behind as a result of time being in the universe at that point. The space that time left behind as space we incorporate as a product when plants assembled structures to form fibre and the material we eat is the time of the plant we eat and the time the animals spent on feeding and to use the time that animals spend on feeding is what we eat to stay alive ourselves. We eat lesser-developed forms of life because we with more developed life can't find the time to accumulate enough heat and carbon and still fill our intellectual life. When you eat sunflower you don't eat sunflower. You use the sunflower plant as a dispensary in which cosmic fuel is stored within the carbon of the plant or meat. By having life the body takes in what other had time to accumulate and then to leave behind. This concept fits everything that forms the concept we have of life notwithstanding whatever rule any one wishes to fit to life.

Life which is what you are, because you are not a decomposable body that the Newtonian intellect describes you as, started accumulating material by thought when you were sperm and egg, and after the Unification of sperm and egg you as a form of life and not a sperm floating around started accumulating useful building material. Take away the thought or life that holds material we see as life and whatever remain destructs automatically because the thought sustaining life and the body life occupies will dysfunction and self-destruct. This is how life as the energy captured in a thought functions…it accumulates carbon to use as building material notwithstanding the form or dimension we think should befit that life form. The instant life leaves the sperm or the body the cosmic material start a decomposing or destructive process that instant. Having life present brings about building material but losing life destroys the fabric that contains or holds life. This is proof that material forms a vehicle by which life moves in this Universe but the body is just a commodity and when life discards the cosmic material it decompose to mere atoms. Life accumulating of material is done by mind controlling the body and I see plants living with a massive intellectual drive.

The plant has the purpose to take carbon from the air then process it by mixing it with heat coming from time or space by the way of frozen heat we call sunlight and build a body we call plants. Then we have life that has the purpose to accumulate carbon from the plants that took carbon from air and use that carbon in plants to fabricate carbon that forms animals. This ability to form the living structure albeit plants or animal is vested in a thought within singularity and singularity is only big enough to hold one thought. Then we humans take animals that accumulated carbon because that is all the animals do all day long and we remove life and use the structure that the previous occupying life-form built in order to build our own body. We use what the animal or plant left behind as time left to form and fill space and apply that then to sustain our life by giving life building material to replace and replenish carbon used in our own body. By devouring what life left behind can we live on what other life left as a result of time spent to form space. Don't allow the atheistic senseless stupidity tell you different. Life is the only energy that can leave the Universe because it is only what life left behind as space filled with carbon that remains as atoms. Life that was never part of this Universe can leave the Universe and leave behind what belongs to the Universe. If your body was what is in charge which is you, then when you are dead someone with life can pump some oxygen into you and shock you with electricity until you bounce around like a ping-pong ball and you will begin life again. Then giving you oxygen and electricity for the brain to function has to replenish the life you had in your body except if life is no more and has gone somewhere beyond this Universe into singularity. That is total rubbish. If life leaves the body there is no structural formation left to control and maintain the structural I integrity of the body. When they shock your heart with electricity it is to get blood flowing and not to put electricity back into your brain. Only blood can do that and to have blood flowing you need life to perform the vital flowing. If you blood stops flowing then you are dead.

Take a close look at a cadaver. It is not something to be scared of because it is a body NOT containing life, as anyone of us will be someday. So it is the same as you being scared of you as you are going to be somewhere in the future and that is pretty silly. On the condition that you were born the only thing you will be someday is dead. If you are alive then you will face death and your only human rite you will ever have is to die and not to vote as politicians shout during an election. What we have to answer is what is the difference between this cadaver in the mortuary and me. One is that the cadaver has no life and I show vital signs filling me with life. The cadaver can't move by it self and I can move by myself. Even if I only have the ability to have my blood to flow it is movement, which is what the cadaver lost. The biggest factor is movement and that movement is linked to thought. Considering the implication of this is vital if you wish to enhance your physical strength and build your body. There are persons in hospital in a coma for years and they apparently show no thought because their muscles don't move and therefore they wither away. The thought gives control over the body and the thought form the muscle and the thought form the size of the muscle. The cadaver or dead person can't get up and walk as I can.

Why can't this dead body get up and walk, it is because the dead has no thoughts. If you think the Newtonian idea is correct that life is part of the body then rethink. I dare you to conduct some tests. If life is electricity as they say it is, then why can you shock that cadaver until it hums like an electric transformer and life will not return? If life is as they say it is electric convulsions then try and shock the brain with electricity and you will find no response. The fact that you can manipulate muscle spasm with electric convulsion shows that life controls the brain by charging electricity and that process is done by thought in life. Life generates electricity that life then implements to control the body life extends for the purpose of serving life. Life is in charge of the body and of thought and not the body being in charge of life. By electrocuting a body with life you merely short circuit life's actions with a stronger jolt of electricity but the electricity is just a modem through which life controls muscles and growth in the body. Then you burn the electricity conducting connection that life has as life controls the functions of the body and do that long enough and life may not find a manner to form conduction of electricity whereby the organ control will become suspended Let us look at the definition of energy. Energy is, as I understand it, indestructible, which means it cannot be destroyed. Energy can only be transferred from one form to another form. At present when energy is measured in work science apparently does not take into account energy losses brought about by anger; fighting and frustration brought about by tribulation but these are as much energy draining as everything else. These are also energy losses and not only work done in the sense of the body being a machine fuelled like all machines. I will have to declare that Newton's statement of energy and work being the same thing is utter nonsense. This is the simplest example we teach children in school and in that they with their utter simplistic views get away with intellectual murder.

Newtonian gossip they call science puts all of life into the realms of the body. Ask what is life and the Newtonian has a lot to offer but put what they offer from an atheist stance and it all falls apart in a sorry way. When does a person die? A person dies when his blood parts the red particles from the white particles where the red particle becomes solid and the white particles become fluid. When being alive the blood flows and that makes the solids interact with the liquids and doing that complies with the Coanda effect on gravity. This meant nothing in thousands of years to the medical paternity but then the Coanda effect still doesn't mean anything to science because science understands nothing about the Coanda effect. Go and try to find an explanation about the Coanda effect (except in my books) and see how far you may get. A person dies when the blood flow stops and the blood stalls because then the Coanda effect using gravitational movement no longer applies as the blood as a liquid does not interact by depositing cosmic heat into the arteries carrying the blood. You don't use oxygen being alive you use the heat oxygen associates with being very volatile and the 6 of carbon with the 8 of oxygen releases the heat to the 7 of nitrogen and then the oxygen going without heat unites with the 6 of carbon to form a unit compound that breathing dispose of. However the collective sense in the maintaining of life sin the supply of cosmic heat to the fibre structure forming the body that life formed.

This process form aging and God knows how Newtonian atheists try to convince the public with a less than honest suggestion that science in the Newtonian manner could in the future keep a person alive for a possible millennium. Eating puts not only cosmic heat into the blood but it supplies carbon fabric to life but also it supplies life-carrying-carbon by which life can replace cosmic tissue. This is one part of life' aging process. Then the next part is the disposing of the replaced carbon 6. Life sustains the body by removing carbon and disposing of it by mixing it with oxygen. The removed carbon that previously formed your body structure as tissue is disposed of through breathing and uniting the removed carbon with two oxygen particles that lost its heat supplement the oxygen as a gas was associated with is the method of doing. In God's creation God never intended man to live forever but to forever become wiser by accumulating knowledge. That is the purpose of life. You learn to live to live to learn and dying is the result of living. By eating you supply carbon with heat as previous life and by breathing you supply heat that burns away old carbon that is removed by paring it off with two part of oxygen for every part in carbon. This is why life ages and this breathing destroy life. I wish to see how they are going to end this process because it is breathing oxygen that supplies heat that allow movement to form life but also aging.

The carbon that replaces the carbon by which the body is renewed and sustains the body in healing and growth is done in a process whereby breathing is inhaling oxygen and exhaling is carbon removed from tissue and then united with two oxygen particles. The oxygen that delivered cosmic heat also removes the used carbon life deposit into the flowing blood. The blood flow is the process of bringing cosmic heat to the body to enable life to apply movement. Life is capable of functioning by movement of a body and movement within a body and any loss of a delivery of heat to the body brings instantaneous death. When oxygen deposited the heat life has to use to move it thereby also remove the used carbon life replaces. Every seven years every atom is replaced in the body. By removing the carbon life rebuilds the body but as gravity changes with the expanding Universe so does the size of the fibre or atoms with which life has the ability of replacing that life has changes to maintain the body and also the fibre size of the atoms that life uses. The body never stops growing and that is why wrinkles form but the body degenerates according to atoms becoming bigger and therefore fewer atoms fit into the original structure. The body outgrows its initial size and thereby collapses under its own growth. It has to perform this way while being part of a body within the growing Universe. This is aging and no medicine can ever stop this process.

This is extremely important to realise that from the first second where life accumulates cosmic material to form life this process of forming life collects tissue that will become a human body. It is not like your halfwit Newtonian professor believes that the human body represents life. From the first moment it is life that forms the body by using cosmic fluid to form movement and it is not the body that forms life. Therefore you with your life forms your human body and it is not your human body that takes the responsibility for life. You are going to age and you are going to die and the only reality in life is death that follows. What is important is not that you have life but it is what you do with life that gives life a purpose.

It is by the thought process that life collects material to form the human body and the human body does not collect life as it goes along. Doing so the ability to do so is the supply of cosmic fluid that is the remaining form that time left behind as space hence space-time or as Kepler put is $a^3 = T^2k$. Everyone in modern science think it is the brain that controls the human body but they are so completely wrong. You use your thought process to control your body and in this thought process you form your body to be as strong as you wish it to be. Even in the very beginning life formed the sperm and the sperm did not represent life because it was life that made the sperm wiggle and if the sperm did not wiggle the sperm was dead or then life-less. You can't have a tube filled with sperm and when you find the lot are dead you then are able to revive the sperm with an electric jolt. You can't have a jar filled with D.N.A and by applying electricity this will regroup the composition and then you build the body of the person once more. You build your body through thinking with your mind. You construct your body cell by cell by using your mind to do so. In the process you take in air that gives you cosmic heat allows you the ability to perform this task. The main issue of life is life needs cosmic fluid to perform as life or welter into death.

How can I prove this? The instant your life vacates the body the body's ability to restructure is gone and the ability to again perform the structuring leaves that very second. The moment life vacates the body, the body degenerates by fragmenting the structure until the entire construction disassembles into forming atoms again. It is life that keeps the body into form and without life the body de-fragments into atoms once more. Your body does not hold life but your mind by thought controls your body and that allows your body life. Your body and life does not run on food and oxygen but on cosmic fluid carried by carbon and oxygen. Life applies gravity to control the body by implementing the Coanda effect, which allows thinking.

In the Universe there is only one fuel and that is cosmic liquid. When anything moves it is because heat is traded for movement. When coal and fossil fuel is burnt the process depends on the release of solar energy stored in the fossil fuel for millions of years. The sun compressed light and the compressing was more than what the sun in its atoms could absorb. By not using all the heat condensed some of the heat escapes into outer space because the space in which the heat is can't manage the heat since the space got cold. A star is not a coal stove but by pumping hydrogen and other gasses on the outside is much more an air-conditioned than a boiler burning coal. When fire burns oxygen it is not the element of oxygen that burns but it is the association it has with heat making the density into a gas that burns when oxygen releases the heat it carries as a gas. If fires did burn oxygen there would be no more oxygen left to burn. The sun does not "burn" hydrogen but pumps hydrogen and by pumping a gas such as hydrogen this action removes heat from the interior to the outside. However it is sunlight that forms the fuel part in fossil fuel as the heat of the sun was trapped in the fossilised remains of carbon life that goes back billions of years. There are three forms of adaptation forming a relation between cosmic substances where there are two substances, one being material and the other being non-materials. When it has no space in-between atoms we think of it as a solid and to be a solid the material must be cold, which means it, lacks heat. When the mixture of heat or non-materials and materials rise we think of the substance forming a liquid. The non-material heat is then more represented in the mixture than was the heat present when it was a solid. When it forms a gas there are much more heat present in the space than there are solid atoms and the density is very low because the non-material substance is overwhelming more. .

Everything in the Universe moves and to be within the Universe forming part of any idea within the Universe there has to be movement going straight and movement going in a circle. That is where we find the essence of Creation as the cosmos informed Kepler mathematically. To understand the cosmos we have to understand why $1 + 1$ is 2 and the cosmos started with 1 growing to 2 and therefore becoming more than what was before. We therefore must know that the following value must be $1 + 1 + 1$ is 3, but why would it be 3. We have to know why $2 + 2$ is 4 and moreover why is $2 \times 2 = 4$. We have to know why $2 + 2 = 4 + 1 = 5$. Understanding this that $2 \times 2 = 4$ and $2 + 2 = 4$ and $4 + 1$ is 5 is the most basic but also the most important aspect of creation. Not knowing why $2 \times 2 \times 2 = 8$ while $2 + 2 + 2 = 6$ indicates a total lack of understanding the dynamics why the cosmos is what it is in all the dimensions it holds.

One must realise that time forms space as "space" forms the history of time left as light in "space" and that the "space" between the earth and the moon is not "space" but it is what time left behind as space. To know the age of the earth and the moon one must take the expanding that happens every year putting "space" between the earth and the moon and put that in relevance to by what time leaves as "space" and from that find the true age of the earth. As the earth "grows" by becoming "bigger" so the moon and earth forms distance by "space" according with time moving. This movement "away from" and "going bigger" is the true and only measure of how much time developed the Universe and that we see as "outer space"

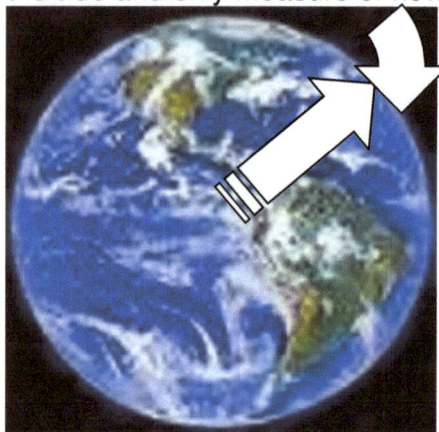

In perspective to singularity relating to the centre of the earth it is the earth that stands still because the connection from the centre to any point on the surface never moves as the line that forms a connection between singularity and the surface remains still. That means all movement is in the liquid aspect. This puts everything that changes as a part of liquid movement although it forms a "mass" connection. Although the line ends at the surface the line running from the centre of the sun is connecting the relevance, which extends to more than a third of the distance going all the way to the next stars, which are Alfa and Beta Century. This line connects every planet to the sun and that is why the Titius Bode law positions all planets. However from the perspective of the sun it is Alfa and Beta Century that moves at a rate the sun can't actually cope with while the sun is standing dead still.

The moon by moving in a twenty-four cycle (according to the earth centre the earth is holding still and the moon is moving in a double shift) as well as forming a cyclic connection of $7° \times 4\Pi^0$ it also rotates once every (about) twenty eight days but it is the moon that moves on both accounts and the earth is dead still.

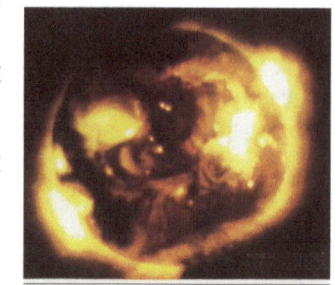

The sun reduces space to a liquid and that the pictures show with liquid and not gas squirting from the star. Instead of looking at what they see they look at the $6500°$ C and according to that scale declare the sun as hot. Hydrogen is a liquid on the surface of the sun and as it comes into contact with outer space the friction caused by the movement makes the hydrogen or some of it overheats again where the hydrogen turns from liquid to gas. The rest of the material that squirts into space does not overheat and returns to the sun as a liquid and as cold as a liquid. The sun moves extremely fast in comparison to all other planets and by moving so fast it freezes the hydrogen in outer space from forming a gas in outer space to becoming liquid within the sun. Gravity is about movement freezing space and we better forget we feel heat and start to think as the cosmos operates. The cosmos is not human and holds no human concepts. Kepler shows in the tables that space a^3 reduces k^{-1} as the sun spins T^2. When space being three-dimensional divides into movement, which is square, space, declines or reduces indicating the relevance (k) goes negative or the distance becomes smaller k^{-1}. This is a mathematical statement showing physics reality that not even Newton can break because this is physics and not that three dimensions a^3 is equal to two dimensions T^2 as Newton stated by

Mercury	$T^2 \div a^3 =$	0.983
Venus	$T^2 \div a^3 =$	0.992
Earth	$T^2 \div a^3 =$	1.000
Mars	$T^2 \div a^3 =$	1.000
Jupiter	$T^2 \div a^3 =$	1.000
Saturn	$T^2 \div a^3 =$	0.999
Uranus	$T^2 \div a^3 =$	1.000
Neptune	$T^2 \div a^3 =$	0.999
Pluto	$T^2 \div a^3 =$	1.004

declaring $a^3 = T^2$. Anyone stating this as accurate has no mathematical sense or has no inclination about sensible physics even if the idiots name is Isaac Newton! The Table I show is Kepler's finding and that shows who is correct, I or science that upholds Newton's views that $a^3 = T^2$.

Individual movement of material occupying specific space is forming density in relevance to all other material moving at various but specific speeds and the faster any atom or material moves, the denser form the movement will make the material to be. Seeing relevancies apply in the picture above it is not the mass that increases but it is the density of the material within the star that increase by claiming less space to hold more material in an denser environment. As cosmic gas or also known as outer space expands it moves slower while the density decreases. The increase of the density of stars reducing space while becoming denser with more material in less space comes about by more material within less space spinning faster because of reduced space bringing about faster circling of material within a smaller confined space. In contras outer space again is moving slower because the space increases through expanding and more space moves slower. This puts the applying relevance on material to move faster in relation to outer space moving slower and thus material becomes denser as it moves faster while it is in ratio with outer space expanding and thus moving slower. This ratio allows material to move faster and then contract more space in the form of heat from outer space, which is filled with non-material heat. As material compact it absorbs heat from outer space that loses density. That secures material growth and by reducing density secures outer space expanding. The star stays the same as outer space expands and that makes that the space the star claims to occupy remained as it was when the Universe in outer space began to expand in terms of the star contracting.

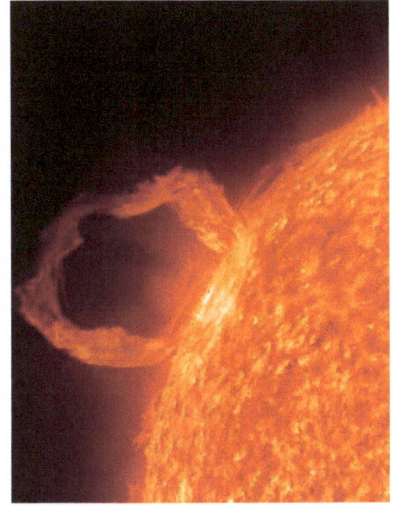

The prominence squirting from the sun can only be liquid heating up as it touches the much more hot cosmic gas. The fact that it rises can only be because it is heating up or getting hotter. If the sun had "pressure" it will release that "pressure" in a cloud of gaseous steam and the sun would go supernova in an instant. The sun spins and this movement contracts the space that by spin or gravity becomes a liquid air that came in as a gas then turns to heat being in a liquid state because of the density increase changing the sun's inner space or atmosphere on the inside from a gas to a fluid that surrounds all the atomic solid particles. Our Earth's atmosphere has all the characteristics of a liquid and is then as such also a liquid and that makes the sun's atmosphere so much denser and therefore so much a denser liquid. At one very specific point gravity compresses outer space from a gas into a liquid and heat in that space then becomes liquid. In the photo's we see the heat returns back to space as it cuts through the sun's curving surface wall bringing about "an explosion". The main issue to realise is that the pumping produces a density increase and the density increase turns the inside from gas to liquid. It is not the oxygen or the hydrogen or whatever fills the container that is a gas or a liquid but it is the amount of space that turns to liquid heat that turns the container from a "planet" into a "star". Even the earth has already some flimsy liquid atmospheres in comparing to outer space. This is the only difference between planets and stars if you insist on having planets and having stars. The sun has no pressure but the excessive movement freezes the gas in the sun into a liquid because the idea of expressing values in terms of temperature is a Newtonian made custom.

When we go in search of what principles applies to form the building material in the Universe we better look and see what is it that the Universe shows us most graphic and we better stop telling the Universe what it is that we want to see and what the Universe should offer us that we wish to see. We better stop telling the Universe it must get mass and start to see what the Universe tells us what it has to offer us to see. If stars burst by releasing heat then stars are constructions that confine heat or cram heat into a small space. If this is true then gravity must be the process of freezing heat by turning movement and displacing space into compacted heat making gravity a process whereby space freezes as it condenses.

When I see a star burst open I should take note of what the star releases and look for the principles applying that should form such a release of heat. Newtonians are forever copying Newton's style by telling the Universe it holds the planets in formational alignment because of their mass while not size nor invented "mass" plays any part in the process. We must stop playing God and create a non-existing Universe and begin to confirm what there is. The biggest concept of being a Newtonian is to tell the Universe what it is instead of looking what the cosmos says told what the cosmos is

When a star burst open it releases massive amounts of heat into outer space. If it exposes heat bursting out then the reason that would apply is it must be because it froze heat into a state of solidity. If the star bursts as it explodes by releasing heat it then clearly overheats. The question never asked is why would stars overheat? We can blame pressure, but pressure would not bring about a star disintegrating from the centre, as the star depicted here clearly does. A burst from pressure should blow the sides out.

When a Super nova goes bang Newtonians say it is because **_"gravity has gone mad"_** and then they still see their position as being intellectual. Since when has gravity got emotions that can go array or "mad". Still more off the point is how can that be to their ability the best answer they can get up to while remaining satisfied with the effort! Stars we call Super Nova has blowouts. It can't be a pressure burst because there is no material wall enveloping the heat and thereby Stars we call Super Nova has blowouts. Pressure release comes from when material containing compressed space bursts its limit. That we humans know since before writing began, but since of late this phenomenon becomes more and more seemingly misunderstood. If stars blow as stars should and as we can clearly see from the picture just above, then the explosion happening to the star we know as a Supernova comes about from other principles, surely.

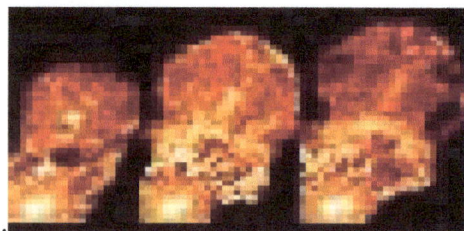

It is very obvious the two occurrences are not a result from the same basic method the Universe uses in destroying stars. When heat surges and becomes too high, it turns into space. That process we call an explosion. It is frequently seen, yet never acknowledged by science. When heat reduces, it relinquishes space in the producing of more concentrated heat, this process we see as cooling.

What ever the terms used there must be a recognising of the inter relation between heat and space where the reducing of the one will lead to the increase of the other. The star does not apply pressure to bring about fusion it freezes the elements into fusion by applying millions and even billions of degrees Celsius. It is our conception of hot and cold bringing total confusion about the principles of cosmology.

I do not think that I or any other person is at liberty to try to calculate any on goings with in the star but from what is clear from the outside one may come to some measured idea of the stars position in space – time. Gravity is the cooling of space by duplicating or moving space, albeit filled or not filled. When the star spins too slowly it does not cool sufficiently and then it becomes warmer inside. As it reaches a point where it overheats because it moves to slow it burst and by expanding the space it regulates the temperature. At a point when it can no longer contain the confined heat it expands and such expanding we call a Supernova occurrence. The contraction of space must be equal to what amount of heat the total number of atoms spinning within the star can retain and gravity is that balance. The fact that it can freeze heat to liquid surrounding hydrogen while holding a temperature of 6500^0 C should be an indication it is not what we seem to acknowledge as normal. The sun is freezing hydrogen to a dense liquid at 6500^0 while space is boiling (expanding through overheating according to the Hubble Constant) at -273^0. Science academics have to review there thoughts on relevancies because what seems to be hot is cold under certain circumstances and what seems to be cold to a point of freezing is boiling hot. There are no standard issue and fit all through out the universe. Every singularity attaché different criteria to borders controlling the space-time with in it rule. What fits humans on earth does not even suit conditions on the moon, yet science cannot appreciate that the moon applies very different standards to that of every structure and every structure is a cosmos on its own turf, supplying its own turf.

We must look at nature to find what is hot and what is cold. Something hot is that which can expand no more because that which is hot expands. When something is cold it can contract no more because it reduced space to the utmost limit.

Outer space is the hottest because it eternally expands while the Black hole is infinitely cold as it contracted what it could contract and keeps on contracting. It is not the specifics that are of importance because the specifics change considerably when taking into account that hydrogen remains in a frozen state at 6500^0 C therefore it is obvious we have to look at other clues to give some indication of what is in process. On earth in the time we have as a duration we find hydrogen freezing at 269 ^{0}C as where it freezes on the sun at 6500^0 C, which implicate the reduction of space to an enormous increase in time duration.

In conditions on earth the rotating velocity of the electron is 3×10^5 km / sec. With conditions being that different it can not nearly be the same in the sun. As space reduces time increases. By having the space reduced to such an extent that it matches near Big Bang relevancies (a period where heat flowed like water and which is the very same conditions we find within the sun) the space would apply accordingly. We also know that relevancies is all about conditions showing similarities under variables and therefore the space and heat component may seem altogether incompatible but is almost the same given the singularity presence within the sun and comparing that to the earth.

What is applying to stars inside the galactica centre is applying to particles inside the sun. Science sees the nuclear reaction but do not recognise and therefore do not admit that the nuclear reaction is three different phases. At the beginning of the process all the heat is solid, placed in a container by nature and the container has a human name called the atom nucleus. In the atomic explosion there are three ingredients that are distinctly apart. When the solid melts down, it becomes a fluid. The fluid we gave the name of light. There is not enough space to explain the detail of the argument, but light is not a gas, it is a fluid. The first step of the nuclear explosion is converting the solid to liquid. In the liquid state the star does not overheat. The overheating becomes part of the second phase. That phase involves the turning of the heat-fluid to a heat-gas we call space. Space is heat overheated creating space, as heat is space concentrated creating a fluid or liquid not yet correctly named.

Every one knows that a gas is one dimension HOTTER than Liquid as liquid again is one dimension HOTTER than being solid. If the star is liquid on the inside, and the liquid evaporates when coming into contact with outer space, then outer space is the hottest, notwithstanding what ever boundaries and values we humans' attaché to the dimension. Our human standards have to change to accommodate the rules layer down by the cosmos and not apply the cosmos to suit our rules of hot and cold, big and small, near and far. In the case of the Super Nova, smoothing prevented the liquid turning into gas, therefore overheating. He liquid froze as a liquid becoming a cosmic lollypop. That which prevents the overheating turned the layers into frozen identities not overheating therefore it became a liquid outside the star. This star was turned into a miniature galactica, sustaining billions of individual singularity, because the governing singularity did not destroy, but the singularity of every nature is still in support of one another. From this picture (and others of Super Nova) one can learn a lot, if one is truly interested in applying cosmic law to the picture and not some human response to what we think would apply to an earth-like star that holds gas as an ingredient. Again I have to press the thought that it is singularity determining space-time form through conditions that bring about the state between matter. Matter can be solid liquid or gas, but it is the condition of the space-time derived from singularity that places the form and conditions valuing the form of the elements. Hydrogen can be as much a solid as gold can be a gas. In the next scenario the overheating core is hotter than outer space and that brings about that the heat will flow to a colder region. In this case the star is overheating and with that can no longer protect its individual singularity. The part the official verdict and mine is in agreement on of creation is that it all started small, but I go one step further by saying the Universe was at the same time eternally large.

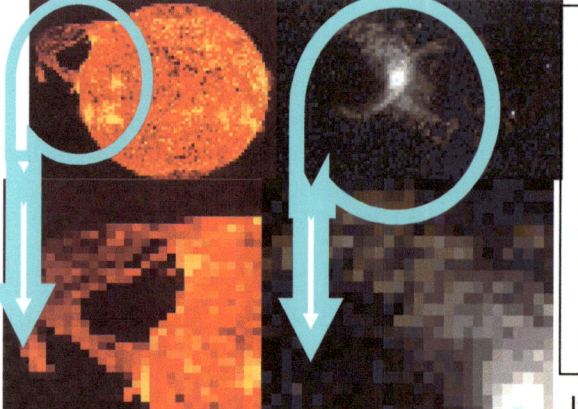

The pictures show clearly the difference of a star NOT overheating being "normal with liquid pouring from it and then becomes a gas as it evaporates. We see the sun exerts heat but still we think of the sun as hot. The sun gets rid of the heat, which means the sun is cold and that is why it removes heat from its surroundings. But because we feel that which the sun rejects we then contribute this heat to the sun by attaching that which the sun removes. It is all about relevancies forming as singularity applying matter in relation to the overheating it started to combat.

It started with singularity producing matter and the matter changed in relevancy to one another by becoming solid or liquid in relation to each other. Space still was at a premium because the space we know and we see as gas, was not yet part of creation. Since the Big Bang the fluid heat is in a process of converting to space enlarging the role of time as the Universe still systematically overheats. The entire purpose of gravity is to combat overheating to allow time growth. From the offset of the first dot dividing as it became the first two dots, it was bringing about the second dot and the eternal number of dots growing from that means that the splitting of the dots assumed as the dots were growing from infinity in size, which is in fact only part of the relevancy because at the same time the infinity presented eternity, where both locked the same value to the dot. This long sentence structure is an effort to explain that everything is linking to another either directly or through other particles and everything came about precisely simultaneously being eternities apart. The stars are in relevancy part of the growing cosmos, where the growing cosmos presents a liquid covering all solid strictures. The structures are no more or less particles irrelevant of size, since time places the value and space is dependent on time. But by the continuing process of the eternal overheating, the geodesic cosmos overheat gradually which presents as the Hubble Constant and this process changes time in space. Since every star holds an individual singularity, separating its singularity from the galactica singularity it is within, it remains as a relative liquid while the cosmos changes its side of the relevancy becoming more a gas.

The difference between the star being the dot we can see in the picture on the outer edge and the star being the dot we cannot see on the inside is the time in promoting the individual singularity. First the star in the centre core changes, starting to collect liquid heat, while the outer part remains part of the cosmic structure. As the Hubble Constant grows the star distinguishes its singularity as it protects the singularity from overheating with the cosmic geodesic space-time. The geodesic space-time is also the outer space, but I prefer not to use outer space. At a point the star becomes a separate structure from the liquid cosmos that turned to gas, and then starts using the liquid to promote the generation of matter in a solid state whereby that matter then later turns to space less singularity as space-time completely breaks down forming neutron stars, pulsars and eventually a completely space less point of singularity in the cosmos being an ancient dot once more we think of by using the name a Black Hole. One has to differentiate between heat and overheating because a star represents the coldest space in the Universe and not the hottest space. Heat and cold are relevant

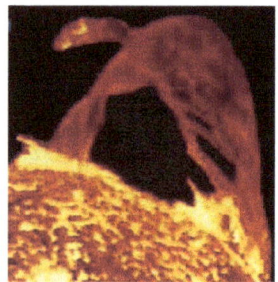

dynamics forming in appreciation of singularity. The sun is the coldest place in the solar system and that is fact. Looking at evidence the sun provides contradict everything science wishes to believe about cold and hot. Science wish to see the cosmos through the eyes of what fits the needs sustaining life on earth and what benefits maintaining surroundings in support of life as one find on earth whereas life has no part in the cosmos except for the speck of dust we call earth. Looking at the cosmos impartial to life the evidence supports another view. Every aspect in the cosmos is the very opposite of what science believe it is. The sun is not a ball of gas but a giant sea of liquid, frozen without any form of gas or air in the interior.

Having a liquid interior the sun has no pressure but has the very opposite of pressure to which there is yet no name given. The liquid comes from singularity freezing space-time within the atmosphere of the sun, and such is the case with all stars still in the shining phase. Stars more developed than the sun is frozen solid causing fusion. Isaac Newton was an alchemist and was not a puritan as Newtonians make him out to be. Isaac Newton did believe in magic but this now is conveyed into the 21st century. The Newtonians measure the surface temperature; test the material on the surface of the sun exactly like the Druids did that came just before Newton did during the Dark Ages they decide that by the magic force of gravity this gas "pulls" into a ball. **Space and heat directly relates being the one form of the other**. As material contracts the space it spins in this absorbs heat by gravitational condensing which cools the material and the size of the material increase. However as material absorbs cosmic liquid the density of the non-material decreases because of a loss of non-material substance going into material substance and materials grow in size while space expands because of loss in density. That is why material grows (the earth seems to grow bigger) and outer space expands (outer space expands) as Hubble indicated. It is all about relevancy changing cosmic dynamics every instant in time. This is why the earth becomes larger and the moon goes "further away from the earth". The density of material increases as the volume size of material grows bigger and as the liquid is concentrated into the spinning object in order to keep it cool by controlling the heat. By removing the liquid from outer space the density of outer space reduces as the heat concentration decreases the relevant space increases by expanding. Material grows by removing heat from outer space to the atom and into singularity as outer space expands by losing the quantity of singularity that concentrates heat. In short it is heat that moves by material concentrating the heat from where the density is less to form more density in materials.

In the picture to the left we find not withstanding whatever name we attach to the red liquid substance flowing from the sun into space and back to the sun, that liquid is heat in a very direct form. If outer space was the coldest place in the solar system the heat should immediately escape to outer space and not return to the sun as it clearly does. If outer space were colder the heat would not return to the sun. All elements forming matter in as much as the heat forming an atom is as much a liquid as it is a gas and a solid. There is no hot as there is no cold. It's about storing energy in space or in heat, which is another cosmic equal being opposing similarities.

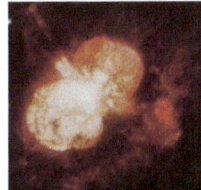

Hot and cold are **relevancies brought about by singularity valuating space-time** and during **the Big Bang** the Universe was **freezing cold** at **three billion degrees C**. It is the relation matter has with heat that provides the form the particle has at that moment. The increasing or decreasing the heat will alter the form of the element. Therefore all elements forming **matter is as much a liquid or not than it is a solid or a gas. It is the space surrounding the atom which provides the form the atom find** its relativity to the rest of the atoms it share space with. **Hydrogen is as much a solid as tungsten is a gas depending on the heat in relation to the space matter is within.** Should **you reply** that it is **the gravity pulling the heat back to the sun,** then that **confirms** my theory that **gravity is all about collecting heat onto matter** with outer space being the hottest place. **It is the concentration of heat in space being relevant to form. When overheating a star turns its liquid to gas whereby it merely transforms its interior to a relevancy it has from pre- to post-Big Bang.** We humans on earth think that hydrogen is a liquid at -259^0 C but that only

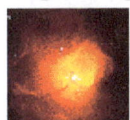apply to the earth. The picture clearly shows the **heat in a liquid** flowing **from the sun** and **back to the sun.** In the **sun the hydrogen holds enormous quantities of heat in a liquid at a temperature of 6500^0 C.** When a star has its singularity secured the star is bitterly cold because it has heat in a liquid

form flowing back to the point of singularity although we may regard the star to be rather on the hot side. The sun (fore instance) freeze hydrogen in a liquid form at 6500 ^{0} C.

If hydrogen remains a liquid at 6500 ^{0} C, just think how cold it must be as the star's interior approaches the point of singularity. Therefore fusing protons comes from cold and not from heat or pressure. By allowing the singularity to overheat the star overheats and heat within the star flows from singularity to outer space freely. In such an event outer space is then colder than the star because the heat releases to outer space with no intention of returning whereas in the sun it returns as soon as it leaves. There are two ways to reduce heat; one is to bring about expanding space, as the photographs clearly show. The second one is where heat will reduce when in motion by spin. When withholding or retarding motion matter will overheat. Gravity is the motion of unoccupied space through the dimensional transformation to occupied space. Motion and space therefore is the anti-, the opposite the negative to heat being the positive. With singularity overheating the expansion of the singularity drives heat into space, creating space to compensate overheating

That is a natural phenomenon. The only reason why **heat will** rather **flow back** to the star than **escape to outer** space once the star released it into outer space is **if outer space presents more heat than**

does the star, because **heat always flows from hot to cold** no matter what influences may arise. Outer space must hold more heat than does the star but the accumulation of space in relation to heat makes it 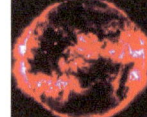 seem colder bringing expanding of heat to become space. The cosmos is all about **converting space to heat** which we see **as gravity** and **returning heat to space** as a **control mechanism** always **keeping** a very delicate **balance** which we see as **a star shining or being normal.** The purpose of the converting of space to heat is to supply the core where singularity is with heat. **It turns space to heat** sustaining matter but sometimes singularity overheats and then matter converts to heat allowing heat to convert to space. That we call many names amongst others exploding into super nova. Whatever the names used is less important 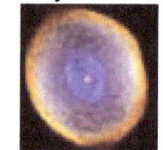 because the process rests on space and heat interacting to form energy. That was what the Big Bang was and the Hubble Constant is all about where matter converts heat to space. I show that space and heat is the very same thing and there is no such a thing as pressure but releasing heat produces space and concentrating heat reduces space with the two interacting on singularity demand setting time to space with time being the spin or motion of heat in space. Heat and space form the second singularity caused by the fragmenting of singularity to compensate overheating during the pre- Big Bang matter forming era. That is what we see as light and space, which again is the same thing and is fragmented singularity forming radiation and heat, where the star re-transfers heat back to space due to an overload.

Since the first instant that time began to convert movement into space the cosmos grew away by forming space and not towards points to destroy space. Material grows and non-material expands. Planets never moved closer, are not moving closer and will never move closer to each other and this is backed by all information collected this past century. **This is how gravity contracts.** In spinning the sphere contracts by measure of 21.991 reducing to 3.1416 while 7 is reducing to form singularity, but also gravity forms when the 7 comes from the past to the present 7 and onto the future 7 and this became 21. Not only that but with singularity advancing from infinity to become one it proves that even as we see singularity as one, singularity also is multi dimensional but that ability is beyond our scope we have being in the Universe. The dimensional change that Π undergoes shows that singularity repeat into a new location by the value of to conform to the roundness of Π changing in value as the circle goes singular. By the revolving of a sphere the space surrounding the sphere compresses as seven changes to one every time twenty-one point nine, nine one becomes three point one four one six. This contraction reduces space in outer space and in that the cosmos grows under a process known as the Titius Bode law. The Universe grows according to the Titius Bode law and not mass and stars condense according to the Coanda effect and not by mass. In this we find the Lagrangian points and the Roche limit forming I limits in this relation between that which expands and that which contracts and in this we have the keys by which the Universe

formed. Gravity is the contraction of space density taking Π from a value of $\frac{21.991}{7} = \Pi$, which is what

is in space to the rim of the earth, which is $\Pi = 3.116 \div \Pi°$. This indicates contraction by the earth's

change in direction by 7° to alter the relevancy applying from $\frac{21.991}{7} = \Pi$ to form $\Pi = 3.116 \div \Pi°$.

The reason why there is something such as gravity is there is a transfer of heat by material movement.

Studying the tables Kepler left us it is very clear that space and not material is moving towards the sun 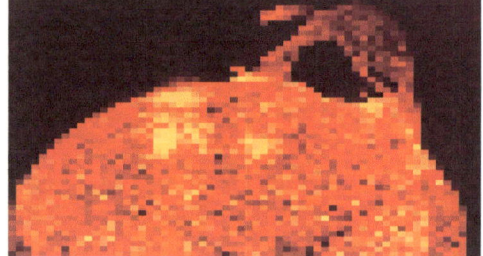 and therefore the sun is contracting space where space is in a process of reducing in volume while the sun is compressing space by a similar margin. Material can't move towards the sun because the Titius Bode law prevents material to come closer to the sun. It is this and three other laws I explain for the first time since the science started. Look very good at every picture of the prominence flowing from inside the sun. What you see is not vapour forming a cloud of mist but a liquid squirting from an even cooler inside. On the inside the movement of the sun freezes the gas forming cosmic gas, which is what is between the earth and the moon to a liquid filling the sun. Again it comes down to enforcing standards our Newtonians apply to life onto the cosmos and the result is stupidity only a Newtonian wise man could be capable of. See the fluid push out of a bowl of liquid, spilling both sides as it falls into liquid. The inside of the sun is not gas but it is fluid. In all of nature there is no NATURAL GAS as much as there is no NATURAL SOLID. Hydrogen is as much a liquid as iron is a gas and neon is a solid. It depends on the element relating to the space/heat in the circumstances surrounding the substance at that very precise instant in time. We have to stop telling the cosmos to show us what we wish to find and start accepting what the cosmos is telling us to find. This shows science must place much less emphasis on life and much more on reality. Life as a comic reality is non-existing and life not withstanding corruption plays no part in the Universe.

Then one final point that would explain the niftiness they craft to protect Newton's fallacies and fables: the formula Einstein introduced being $E = mC^2$ is that that original and did not that great and is even less admirable seeing it is the formula Kepler introduced and which Newton raped because Newton never had the brains to evaluate and understand the formula Kepler gave to the world. It is just the first truthful formula that applied meaningfully and science could work with seeing nobody ever made the effort to investigate Kepler's formula and I would think Einstein did it for the first time. But if he did he never gave credit to where credit is due. The formula $E = mC^2$ should read $E^3 = mC^2$ and because if E is equal to m^1 and C^2 then E must exponentially be $^{1+2=3}$ or then E^3 which in full is $E^3 = mC^2$ and in that it is that is a copy of Kepler's formula $a^3 = kT^2$. Now today it is claimed that this Einstein formula $E = mC^2$ opened the Universe to the understanding of man and all that shouting is a deliberate sidelining of the truth. The truth is that Newton's claims on Kepler's work is a hoax and to divert the attention away from this fact everyone gave great credence to Einstein formulating nuclear power as $E = mC^2$ and all the while this formula dates back directly to Kepler and Kepler's formula the cosmos gave him as $a^3 = kT^2$.

Mercury	$T^2 \div a^3 =$	0.983
Venus	$T^2 \div a^3 =$	0.992
Earth	$T^2 \div a^3 =$	1.000
Mars	$T^2 \div a^3 =$	1.000
Jupiter	$T^2 \div a^3 =$	1.000
Saturn	$T^2 \div a^3 =$	0.999
Uranus	$T^2 \div a^3 =$	1.000
Neptune	$T^2 \div a^3 =$	0.999
Pluto	$T^2 \div a^3 =$	1.004

Then the only great thing about Einstein's formula is that science managed to cover the truth bout $a^3 = kT^2$ being the same as the Einstein breakthrough of $E = mC^2$ that must be $E^3 = mC^2$ and is just Kepler's $a^3 = kT^2$. It is in small detail that the lie covers the truth by blind sighting people as some manor detail misleads the correctness covering a lie to protect the cover-up they have to protect to keep them in office and get the billions of funding rolling in.

By positioning the point where I prove singularity is I managed to prove many aspects in cosmology that is still unclear or not understood. I prove and formulate the Roche limit as having two factors, both of which play a most dynamic role in the cosmos. I have also managed to prove, formulate and define the Titius Bode principle and in that principle also comes with two factors. The principle is a derogative of the Roche limit and in amongst that there is another principle I have discovered concerning the dynamics as well as the role of light in the universe. The Lagrangian principle also flows directly from the Roche limit as the Roche limit is a ratio in conjunction to the point which I claim singularity is. This is the normal manner in which science presents the solar system and this is as false as the value of money. Look at the way the planets are arranged and then see how it is possible that "mass" can put the position of planets in accordance to the way the planets are arranged. How can science in the light of this evidence still maintain planets are positioned according to "mass"?

To intimidate and discourage people science places formulas that have no purpose at all but to put the fear of God into in place to prove…nothing because they and nobody else can use these meaningless formulas. Now the question is that if the "mass" was pulling the comet closer, what then is pushing the comet further away. Never do I get any answers about these and many more matters science can't explain and yet they teach the children that "mass" pulls "mass" and to that effect there is no evidence proving this statement. There is no way in hell that any person can use the mass of Jupiter to pinpoint its position in the Universe and all these formulas this use is total criminality because the formulas are nothing more than the cover –up to conceal the fact they have no idea why the solar system and indeed the entire Universe is what it is. This book is the first book in the history of man that will explain why the solar system and in fact the Universe form in the manner of the Titius Bode law. Getting away with all the deception they scare people witless with the most impressive mathematics they can think of so please when you see the mathematical formulas I introduce don't let it scare you because they're using of it is only to scare you and that is all.

Any person that shows interest in investing in the promotion of a book in whatever segment or category, contact me at any of the following e-mail addresses and tell me which category would your interest be. www.questionablescience.net http://www.singularityrelavancy.com www.sirnewtonsfraud.com and then arrange to have literature sent to your address. The reading is comprehensive if any person should feel to read all or many of the books on offer and the topic requires concentration except in the commercially written category I labelled a price in the order of $6.00. The other categories are also available going at $10.00 but please keep in mind that the reading requires better concentration because the reading seems more complex ad the more technical the explaining gets the fewer books in volume would sell so the turnover will be less. The example I now present is one aiming at the higher end which I would sell at $15.00 giving each partner the author, the publisher and the promoter about $5.00 per sale.

Any person who shows interest in investing in the promotion of a book in whatever segment or category, contact me at any of the following e-mail addresses and tell me which category would your interest be. Interested persons can also go to Lulu.com and download material (there are some for free) or contact me directly to find more information about books I offer in the intended partnership. My private e-mail address is or the websites are: mailto:orders@sirnewtonsfraud.com .
www.questionablescience.net   http://www.singularityrelavancy.com/   www.sirnewtonsfraud.com

Yours truly

P.S.J. (Peet) Schutte.
Some of my websites are:
www.questionablescience.net   /   NaturesCosmiConcept-E-Z-R   /   Titius-Bode-Law-Explained   /
NaturesCosmiConcept-E-Z / www.sirnewtonsfraud.com / www.questioneblescience.net

Revealing a Science Cover-Up

ISBN 978-1-920430-58-0   Written by P.S.J. (Peet) Schutte
© KOSMOLOGIESE EN ASTRONOMIESE TEGNIKA

All rights are reserved.
No part, parts or the entirety of this book may be reproduced by publishing, electronically copied, duplicated by whatever means that form reproduction or duplication, without the prior written consent of the copy rite owner.

This is the book showing everyone that there is A Conspiracy in Science in Progress I honestly started this book 52 times, each with a unique and different start because as I wrote on and started to explain something I knew was new to a reader, then as I got to a point where every time I touched another aspect it was a critical aspect of information needed to fully understand what I tried to say. I found some information that I first have to share before what I shared before would be understood by any reader and I found myself deciding I have to start with that…until I came to another aspect and at that point I realised this was more important than what I first started with after I started with the one just before the one I started with and by the time I came upon the sixth start I realised the first was even more important than was the seventh. You see what you are about to read is very much as new as anything new you will ever come across. You are the first to read what you are going to read for the first time. Now I give you many ways to approach the information you are to read (for the very first time) and I now decided to leave it up to your decision where you would like to start. You can start at any introduction and read the part you decide on first. I leave you this option because I have no idea what people may deem as to be complicated and what people may find easy reading…to me it is all the same because I have been living with all this information you are about to read and then about ten times more for most of my adult life. However, there might be those that think what you read is simple but for those I say: believe me this book does not even form the introduction considering what information the rest of my work presents. I have endured criticism by persons that clearly have the science insight of a skunk and the understanding about science of a porcupine and they would not be able to read one paragraph of my Theses I named **_Matter's Time In Space: The Theses_**. If anyone wishes to criticize me you are welcome but present facts and not an idiot's slurring. Why this response: It is your attitude that helps to keep cheaters in office conspiring to keep the human minds unaware of a very important truth they successfully hide. No matter how informed you think you are, if you aren't able to explain how the cosmos started, not from the simplistic rhetoric Big Bang, but from where the first dot became the second dot and then the third dot, and why the first dot became the second dot and then the third dot then just remember I can and I did. In some of my books I explain how this happened. I will leave you a clue; it is because 1 and 2 is 3 and why $2 + 2 = 4$ while $2 \times 2 = 4$. This configuration in why mathematics is a line using numbers might seem simple, but that is how it all started and it's more complicated than anything you think of as being science.

I know I have to keep a very complicated subject very simple to draw as many readers as my abilities possibly can muster and I have to make as many issues about what forms our Universe as little complicated so that the message is as broad as it could be and tell as much as I will get away with while still remaining as coherent as I can in order to enable as many persons as what I possibly can to understand never heard before proof about Creation and moreover about how there can be a concept science has been more than dishonest about and this dishonesty I have to convey as simple but also as honest as my meagre abilities would allow. If you can follow this introduction, then go ahead there are much more to-be-understood- complexities in the offering. The essence of all of this is that everything about this book has never been told and when I introduce any aspect, it relates to something I have not yet mentioned and then that requires explaining and then that explaining requires a lot of introducing of supporting facts. On something as simple as a top spinning I wrote one book (not this book) explaining the cosmic implications by what is a simple toy yet can spin becoming the same as a cosmic star. Then on to top this I have to prove what is untrue when knowing persons were brainwashed in believing science is truer than any religion ever could be and to think this way they were hoaxed by tutors in the past making everyone trust in a religiosity without contemplating flaws hidden science. By the top performing spin it is apparent that all we as humans can do is to manipulate cosmic behaviour. The question is what is cosmic behaviour? When you reach the final page and only after you completed this book will you know a very small part of what is authentic cosmic behaviour because this is the first time ever in all of history that the cosmic pillars are introduced? If you study a toy top spinning you will find true cosmic behaviour. Why is that? It is because we humans can create nothing but mimic cosmic behaviour.

Can you raed this beuseae olny srmat poelpe can and to raed and aulaclty uesndnatrd this you hvae to be srmat. If you can raed this slpeling you psas one prat of my raednig tsest. I cdnuolt blveiee that I cluod aulaclty uesndnatrd that I was rdanieg what I was wrttign. This tsets the phaonmeal pweor of the hmuanmnid and aoccdrnig to a rscheearch at Cmabrigde Uinervtisy, it deosn't mttaer in what predr the ltters in a word are, the only iprmoattnt thing isthat the frist and lsat ltteer be in the rghit pclae. The rest can a taotl mses and you can still raed it wouthit a porbelm. This is beuseae the human mind deos not raed ervy lteter by istlef, but the word as a wlohe. Amzanig huh? Yach and I awlyas tghouhot slpeling was ipmotantt!

7H15 M3554G3 53RV35 7O PR0V3 H0W 0UR M1ND5 C4N D0 4M4Z1NG 7H1NG5! 1MPR3551V3 7H1NG5! 1N 7H3 B3G1NN1NG 17 WA5 H4RD BU7 N0W, 0N 7H15 LIN3 Y0UR M1ND 1S R34D1NG 17 4U70M471C4LLY W17H 0U7 3V3N 7H1NK1NG 4B0U7 17, B3 PROUD! 7H15 PR0V35 0NLY C3R741N P30PL3 C4N R3AD 7H15.

  PL3453 REG4RD 7H15 M3554G3 1F U C4N R34D 7H15 :)..

If any person can read the above, then that person must have the intellect to figure out and see what fits where and how things fit and do not fit. What is above my understanding is that intellectual people can't see what is wrong with science when I show exactly the incorrectness science has

You are going to read some mathematics in equations and expressions in mathematical formulas placed to defend my position but if you don't like it then just skip the mathematics because the content and grounds the mathematics proves or disproves is not important in the arguments and it is there for physicist to hide behind. I don't have to hide behind mathematics to make others feel inferior because my arguments make people understand physics and make people feel empowered and superior. The mathematics I include is to show what mindless clots those Superior Humans are that portray their position as superior in mathematical ability and it is there to disprove the Members of the Physics establishment that advocates the necessity of bringing mathematical proof to prove? It is there not to scare readers away but to silence the Brainy Bunch critics by showing them the foolishness of their arguments. By using mathematics the Brainy Bunch have been cheating the public and have been brainwashing students for centuries. That cheating is how they do it and I have to show and uncover the dishonesty in mathematics. However, IT IS NOT important to understand the formula in any way, so glance at it if you wish or skip it completely because if you do not wish then it does not concern you personally! The mathematics are a smoke screen in place to scare the daylights out of anyone not schooled in the art of mathematics and to impress those that understand mathematics as well as cover the truth about the fable everyone in physics pretend to guard. So skip it because it protects a fable.

Just to inform the commercial readers: this book doubles as a manuscript that will serve to be handed in at a University to be adjudged as a New Theoretical Cosmic Concept.

To Academics this: As I am writing the opening prologue, I am well aware that I have no hope that my effort in sending you a copy of this new cosmic concept would urge Academics to change science as science should change. I am aware that sending the full manuscript will not ensure the reading this book because this is not for the self-opinionated and those that can't form concepts outside what they were taught at school. When you have the mindset to read this book from start to finish then that would then be the only way to entice you to form well opinionated response. Believe me I have in time since 2000 sent so many attempts to various academics and many institutions in the past and never yet had I received any response, albeit positive or negative...and therefore I am under no illusion that it would happen this time where I will find science is changing to evict backward philosophies and outdated principles that those in science hides from detection. I just can't see that any academic will come forward and admit science is wrong and thereby throw away the years of devotion that they enjoy as being the most brilliant minds on earth. However, without my trying, thereby not even trying, I then forfeit even the least of any sort of chance I would ever have of finding a positive response from any institution albeit by a long shot.

My work is in every detail undoubtedly correct just as science should be. Therefore, even however slim the chance might be that there might be any academic response to this work and my doubts I have about the world of science reacting in a meaningful way, then by not trying yet again, I deny myself of having any chance I might ever have in finding justice for my life-long battle. That means although I know I have no hope in charging a response from the academic world of astrophysics, when not publishing the work, I will change the uncertainty of having no hope to a very certainty of finding no response from any quarters ever. I have no chance on success, but while sending this manuscript out where there might be some intellectual responding with vigour, in my attempt I still do have hope while if I do not send you my work, I will not even have a chance on hoping... and with that in mind I am sending a complete manuscript that covers a lot of supporting facts. I know this manuscript is intense but if the work is not entirely appreciated my having you read an entire manuscript with much detail will bring doubt and then I have no hope.

The theory that I present to you is conclusive as it is totally comprehensive and is covering cosmology in as full content as anything ever did before. It is not the facts about my work that seems odd, but it is the breaking of century's worth of misplaced culture that makes my work seem absurd. My work in overview becomes a big jigsaw puzzle and without the entire puzzle placed on the table, my work would not make sense, and I am the first to realise this matter! The volume of information that my work undertakes in changing science can't fit in a journal because even with what little information this book holds alone about the overview of my work in entirety, what this book holds will change every aspect currently understood by science, and when I say that, it is not a madman boasting, for I challenge any person that are able to read this work, to show that my statement of changes that has to come to physics are not true.

However, my claim about this work having the ability to bring change to science cannot come from some brief communication about a few topics evaluated by the way of glancing over some aspects. Changes don't happen when I lightly touch on a number of thoughts to be studied on the incorrectness as to how science thinks of gravity as a force of attraction. Changing the fundamentals of science would entail a massive undertaking, covering a comprehensive background of details and must result from a study of gigantic proportions. This is what my work entails. It is requesting changes of fundamentals about the entirety of science. There is no chance in hell that the content containing the fundamental changes required to correct physics as a whole can be dealt with by simply reducing the information that just this book alone presents as little as it is, to be contained by an article presented in a journal. If it is thought that the information presented in the other twenty or more books will also be added to the information in this book, the entire thought of confining the concept to an article in a journal becomes bizarre.

This statement I make about me being able to understand cosmology could only become true and sound in the fundamental covering of the concept, when the entirety of the information that is presented in this book alone, is fully covered by all the detail I present to you. Leaving out even one thought that may be viewed on its own as insignificant when standing apart from all the rest of the other thoughts, be it as fundamental as such a thought might be, then the exclusion of that one thought will possibly nullify a vital part of the entire concept as a whole that I present about the facts that the view that should be changed. When leaving out one argument without the support of the rest of the content in this book, in doing so it will then sound like the delusions of a drug addict raving incoherent inconsistencies and would then not find support as the compliment of arguments support each other. Without reading the rest of the information, anything you are then about to read, without all facts supporting the entire issue will sound trivial. Here are a few thoughts about the new cosmic concept but what you read remains a small part.

Again I repeat: You are going to read some mathematics in equations and expressions in mathematical formulas placed to defend my position but if you don't like it then just skip the mathematics because the content and grounds the mathematics proves or disproves is not important in the arguments and it is there for physicist to hide behind. I don't have to hide behind mathematics to make others feel inferior because my arguments make people understand physics and make people feel empowered and superior. The mathematics I include is to show what mindless clots those Superior Humans are that portray their position as superior in mathematical ability and it is there to disprove the Members of the Physics establishment that advocates the necessity of bringing mathematical proof to prove? It is there not to scare readers away but to silence the Brainy Bunch critics by showing them the foolishness of their arguments. By using mathematics the Brainy Bunch have been cheating the public and have been brainwashing students for centuries. That cheating is how they do it and I have to show and uncover the dishonesty in mathematics.

However, IT IS NOT important to understand the formula in any way, so glance at it if you wish or skip it completely because if you do not wish then it does not concern you personally!

The mathematics are a smoke screen in place to scare the daylights out of anyone not schooled in the art of mathematics and to impress those that understand mathematics as well as cover the truth about the fable everyone in physics pretend to guard.

So skip it because it protects a fable. Everything they calculate is pure fantasy that they wish you to believe because they say so…and that they do because they wish you to feel inferior in contrast to how they see their position as entirely superior to all. They are living a lie you allow them to live.

I include the mathematics to silence simpleminded yet opinionated because although they have no idea what I am talking about, yet they feel a desire to criticize me about how I introduce science. All those that criticize me about why I criticize science should rather look at what I criticize science about and not their admiration about the cultural To them and to those that can't even understand their own simple ness should you feel the need to criticize me then do so on the correctness of the Cosmic calendar Layout.

The truth about a conspiracy is that everyone involved with the conspiracy will fight tooth and nail to stop the conspiracy to get leaked and leave those behind the conspiracy and all those feeding from the corruption of the conspiracy exposed.

Those that have the power to maintain the conspiracy would keep the waters as still as possible as to draw no attention to such conspiracy. A true conspiracy has to be as quiet and as unseen as it could be. A true conspiracy must involve everyone without anyone detecting even a hint of what the conspirers hide. There are many conspiracies going on such as the banks involvement with crime and the bankers profiting from gamble rackets and drug selling. The same goes for the Insurance business profiting from lenient sentencing of courts holding very merciful judges in office so that the crime cases and the burglaries, car theft, hi-jacking and all other forms of crime will shoot through the roof every year and grow by thousand percentage points from which insurance will sell ever-more cover-policies.

The more crime is about and committed daily, the more people need insurance covering and thereby the more money insurers' bank giving bankers profit to spend on the stock exchange by controlling the economy. Bankers buy democracies with money they give politicians to write laws that protect the rich against the poor. In time I might write about this but now I cover another conspiracy, a bigger case file about a conspiracy everyone on earth participates in…it is about…

The Universe consists of gravity that forms by the working of the four phenomena never mentioned.

The Titius Bode law has been around for centuries and with all the mathematical splendour available there for all to use, all the brilliant mathematicians could never come close to show any ability to any understanding of this very important phenomena. They could mathematically equate the formula the sequence applying as the formula, but then after that their superior human intellect dries up.

The Roche limit has been around for centuries and with all the mathematical splendour available to apply in order to fathom concepts behind this phenomenon, still with all the computing ability of a machine all those physicists with all the mathematical superiority could not touch any understanding about the concept forming the background. Yet hen using the truth in physics the answer is simple.

The Lagrangian points has been known to science for centuries and with all the mathematical splendour available not one calculation could ever explain why this event is taking place.

The Coanda effect has powered turbine engines and aeroplanes in flight for almost a century and with all the mathematical splendour available to design the most terrific aircraft, not one engineer could mathematically compute one fact to show understanding why this takes place. How sad it is that those claiming of much superior intellect in physics remain just no more than having computing power.

That does not say much for the bountiful prestige that mathematician's claim as their lawful bragging rights in areas where true human intellect is called on. Is it not high time to begin to admit you are playing the game of fools with you arrogance about your achievements using mathematics when designing space whirls and travelling to galactica while not even understanding what movement asks for? You do not even understand the neutron and the neutron is compressing density increasing, which is what gravity is, which is what time is, which is what all movement is…that is why the neutron has no mass because mass is the principle coming about where independent movement ends.

You're mathematics could not get you any closer than playing games in a fairy tale Universe using misguided presumptions about mass forming gravity and living the Universal farce which Newton created because that fairy land is what all the Kings clever heroes and all the King's splendid wise could never prove in hundreds of years. If you feel superior as a scientist practising physics on the highest level having a gloating hail of superior mental capability covering you like an aura, then I have very saddening news for you. If you have the ability to compute and calculate at the highest level, then look at your computer and see one that machine has abilities as a machine which is equal to you, but it's a manmade machine. Stop playing games by creating fairy worlds making up fairy tales about fairies and little people, mass that can create forces, four of them no less, and come and join the rest of us living in reality that does not need to compute forces to be able to not understand what it is that you compute, but to use human intelligence and in that way to understand what only human intellect could ever understand. Then what in the present is not worth carrying into the future as the past being worthwhile?

Dual singularity DISLIKES this book
"Anti-intellectualism has been a constant thread winding its way through our political and cultural life, nurtured by the false notion that democracy means that 'my ignorance is just as good as your knowledge.'" Isaac Asimov

Vern11ricia DISLIKES this book
Science is not black and white. This is an extremist view of personal opinion with little substance. Proof of this is that most great scientist were or are very religious people who strove for understanding of complex subjects which many of us do not care to understand.

Larz Law DISLIKES this book
This book is in concise and displays no knowledge about what is being argued. The Books first page is grossly wrong and trying to say gravity is wrong is the worst way to start a book like this. I am as of yet undecided as to my view of the world and what forces are at play (divine or other wise)

Those that feel it their task to judge me as they are opinionated about information they were never able to read let alone understand, first explain to me where I am wrong in my relevancy about how the cosmos came about. This is a colander that the cosmos follows by which it developed to where it now is. Go on and criticize me but not with slogans and misconceptions you have been brainwashed to believe. Try to use arguments, facts and concepts, it shows intellect and not the zombie-minded state of the critics above. I flourish on your wisdom and your insight drives me on to the next day because in that we have the only valid reason why we are alive! I want to know what others know so that I could be more fulfilled. Can you believe some one with a single brain cell accuses me of "Anti-intellectualism". In all my life I have never come across a fool such as that and still this idiot is stupid enough to be able to think he or she is worthy of an opinion? You mindless slob, you have not the brains to understand true science.

I am all for critics but do it with intelligence. Show me that the solar system does use mass and that there is a reason why the biggest planet is in the centre and the smallest planets are on both ends and that it is because mass does apply! I believe we on earth have one purpose and that is to learn. We live to learn because we learn to live and the day any person learned nothing that day that person died. There is one reason why we breathe and that is to accumulate knowledge and assemble wisdom.

The Cosmic Calendar showing the route the Universe follows as it followed the path of Creation.

This is a relevancy indicating the development line Creation follows

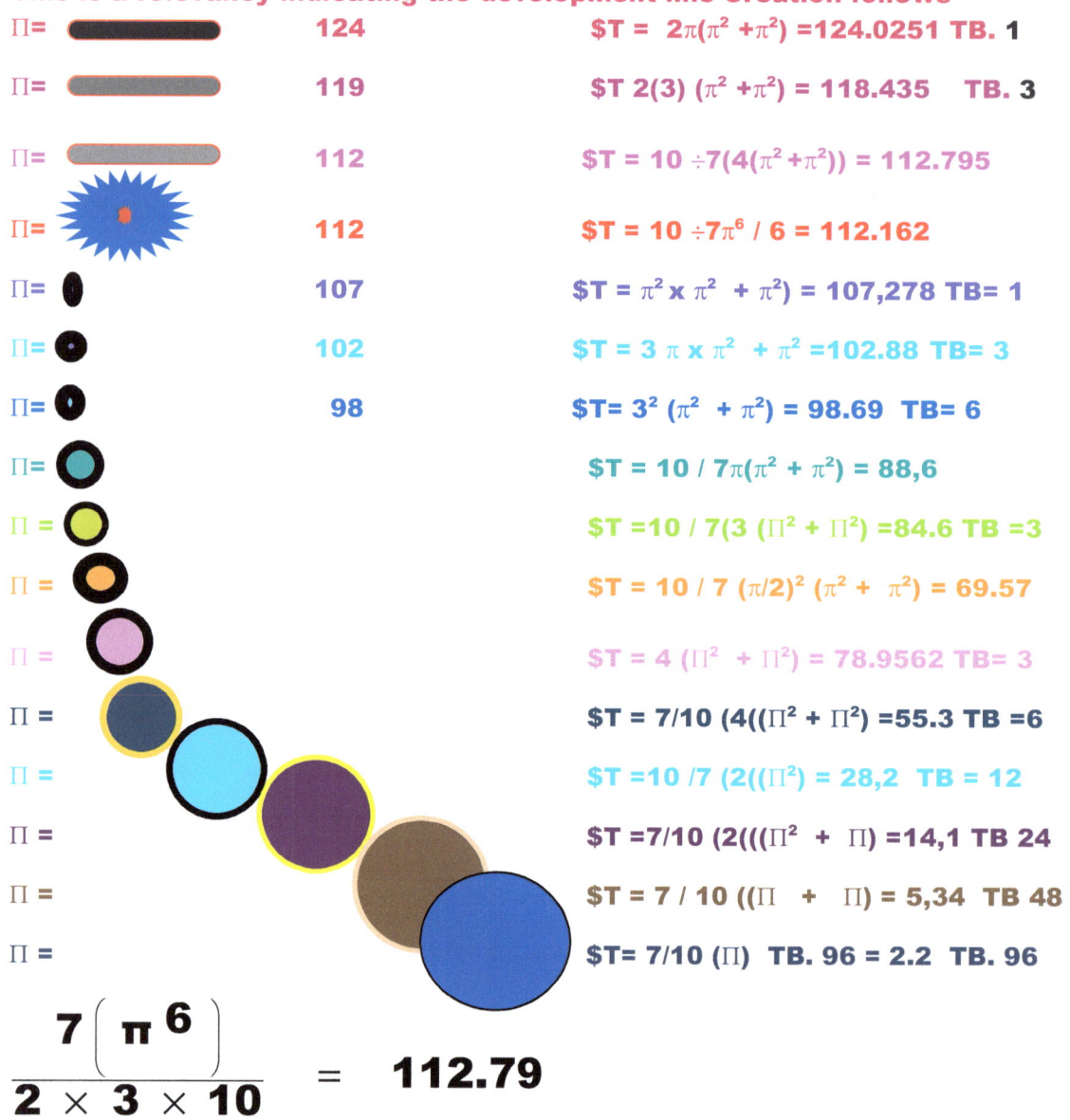

Π=	124	$T = 2\pi(\pi^2 + \pi^2) = 124.0251$ TB. 1
Π=	119	$T\ 2(3)\ (\pi^2 + \pi^2) = 118.435$ TB. 3
Π=	112	$T = 10 \div 7(4(\pi^2 + \pi^2)) = 112.795$
Π=	112	$T = 10 \div 7\pi^6 / 6 = 112.162$
Π=	107	$T = \pi^2 \times \pi^2 + \pi^2) = 107,278$ TB= 1
Π=	102	$T = 3\ \pi \times \pi^2 + \pi^2 = 102.88$ TB= 3
Π=	98	$T = 3^2 (\pi^2 + \pi^2) = 98.69$ TB= 6
Π=		$T = 10 / 7\pi(\pi^2 + \pi^2) = 88,6$
Π =		$T = 10 / 7(3\ (\Pi^2 + \Pi^2) = 84.6$ TB =3
Π =		$T = 10 / 7\ (\pi/2)^2\ (\pi^2 + \pi^2) = 69.57$
Π =		$T = 4\ (\Pi^2 + \Pi^2) = 78.9562$ TB= 3
Π =		$T = 7/10\ (4((\Pi^2 + \Pi^2) = 55.3$ TB =6
Π =		$T = 10 /7\ (2((\Pi^2) = 28,2$ TB = 12
Π =		$T = 7/10\ (2(((\Pi^2 + \Pi) = 14,1$ TB 24
Π =		$T = 7 / 10\ ((\Pi + \Pi) = 5,34$ TB 48
Π =		$T = 7/10\ (\Pi)$ TB. 96 = 2.2 TB. 96

$$\frac{7\left(\pi^6\right)}{2 \times 3 \times 10} = 112.79$$

SPACE-TIME RECEIVES A SEPARATE VALUE AND THE COSMOS GETS DIMENSIONAL

In the event of any readers who may have questions concerning more facts as it is presented in this book; please feel free to contact me, PEET SCHUTTE. All information divulged came about through independent self-study during the past forty-two years or so.

I have to warn the readers that the topics are showing a very new approach with no quick answers abstaining from proof or holding just a few lines and the information is new in nature but not hard to grasp. Should anyone wish to confront me or wish to contact me then do so by E-MAIL AT: e-mail
mailto:info@questionablescience.net
Yours truly

P.S.J. (Peet) Schutte.

www.ingramcontent.com/pod-product-compliance
Lightning Source LLC
Chambersburg PA
CBHW050724180526
45159CB00003B/1119